Food Enzymes

The Missing Link to Radiant Health

Food Enzymes

The Missing Link to Radiant Health

Humbart Santillo, B.S., M.H.

Edited by Deborah Kantor

HOHM PRESS

Hohm Press, Box 2501, Prescott, Arizona 86302

First edition published in 1987.
Second edition 1993.

Library of Congress Card No. 93-078887
ISBN 0-934252-40-8 (pbk.)

The theories and formulas presented in this book are expressed as the author's opinion and as such are not meant to be used to diagnose, prescribe, or to administer in any manner to any physical ailments. In any matters related to yoour health, pleae contact a qualified, licensed Health practitioner.

DEDICATION

To:
Tony Collier, National Enzyme Company, and to
Lou Piccone, N.F.L. alumnus

When a man needs answers, he must reach out
and seek from those who are closest to him.
These men provided me with that avenue.
Thank you.

TABLE OF CONTENTS

INTRODUCTION

Throughout my years of experience as a health practitioner, author, and researcher, I developed a certain understanding: there is no singular formula, food, or health product that is a "cure-all." Even though I intellectually felt this to be true, I firmly believed that there was something that every person could use during therapy, or as a health supplement, that could act as foundation and adjunct to both medical and nonmedical therapies. Now I feel my wish has come true. I found that what I was searching for was food enzymes.

If you are interested in longevity, vitality, superior health, overcoming sickness, or if you are having trouble losing weight and feel that after taking vitamins and minerals for years you haven't really benefited as much as you would like to, this book should be of interest to you.

Three years ago, Viktoras Kulvinskas, author of *Survival into the 21st Century,* sent me a book written by Dr. Edward Howell called, *Food Enzymes for Health and Longevity.* This book explained to me why some therapies work and why some don't—because of enzymes. Enzymes are needed for every chemical action and reaction in the body. Our organs, tisues, and cells are all run by metabolic enzymes. Minerals, vitamins, and hormones need enzymes to be present in order to do their work properly. Enzymes are the labor force of the body.

WHAT AN ENZYME IS

In 1966, a Scottish medical journal stated, "Each of us, as with all living organisms, could be regarded as an orderly, integrated succession of enzyme reactions."

It has been felt that enzymes are protein molecules. This is incorrect. Let me clarify this by giving you an example: a light bulb can only light up when you put an electric current through it. It is animated by electricity. The current is the life force of the bulb. Without electricity we would have no light, just a light bulb, a physical object without light. So, we can say that the light bulb actually has a dual nature: a physical structure, and a nonphysical electrical force that expresses and manifests through the bulb. The same situation exists when trying to describe what an enzyme is within our body structure.

Let me try to define "enzymes" and a few properties about them. An "enzyme" is said to be a protein molecule and each acts in certain ways in the body doing specific jobs such as digesting food, building protein in the bones and skin and aiding detoxification, to name a few.

Once we cook food at high temperatures, however, the enzyme is destroyed. It no longer carries on its designated function. Although the physical protein molecule is still present, it has lost its life force. Much like a battery that has lost its power, the physical structure remains but the electrical energy which once animated it is no longer present.

A protein molecule is actually only the carrier of enzyme activity. In experiments described in *Chemical Reviews* (1933), the activity of one protein molecule was transferred over to another protein substance, leaving the original molecule devoid of its original activity.[1] This only proves further that an enzyme is the invisible activity or energy factor and not just the protein molecule itself. So, for clarity, let us agree that a protein molecule is a carrier of the enzyme activity, much like the light bulb is the carrier for an electrical current.

WHAT DO ENZYMES DO IN THE BODY?

Enzymes are involved in every process of the body. Life could not exist without them. Enzymes digest all of our food and make it into a form small enough to pass through the minute pores of the intestines into the blood. Enzymes in the blood take prepared, digested food and build it into muscles, nerves, blood and glands. They assist in storing sugar in the liver and muscles, and turn fat into fatty tissue.

Enzymes aid in the formation of urea, which is to be eliminated as urine and also in the elimination of carbon dioxide in the lungs.

There is an enzyme that builds phosphorus into bone and nerve tissue, and another to help attach iron to red blood cells.

Male sperm carry enzymes that dissolve the tiny crevices in the female egg membranes, so that a sperm may gain entrance into the ovum.

An enzyme called "streptokinase" is used in medicine to dissolve blood clots. Enzymes in our immunity system attack waste materials and poisons in the blood and tissues. These few examples exemplify the importance of enzymes to our everyday body functions, Howell IX.[2]

The number of enzymes in the body is overwhelming, yet each one has a specific function. A protein digestive enzyme will not digest a fat, a fat digestive enzyme will not digest a starch. This is frequently called "enzyme specificity." Let us say enzymes are very intelligent when it comes to their activity and functions.

Enzymes act upon substances and change them into another substance, either chemical or a type of by-product, but remain unchanged themselves. Any substance that an enzyme acts upon is called a "substrate." The substrate is then changed from its original identity by the enzyme to another substance with a different identity. Each enzyme is believed to fit into a specific geometrical design as shown in Diagram 1 on the next page. Enzymes do a tremendous amount of work.

DIAGRAM 1

HOW AN ENZYME WORKS

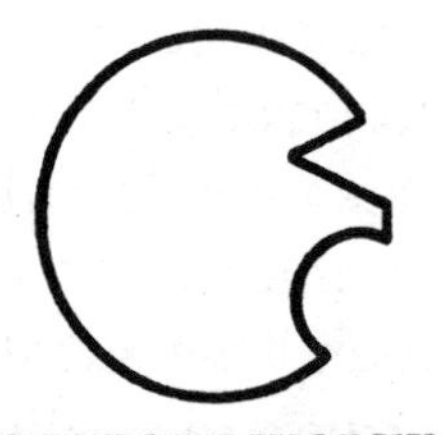

ENZYME BEFORE WORKING

MOLECULE COMPLETE

ENZYME WHILE WORKING

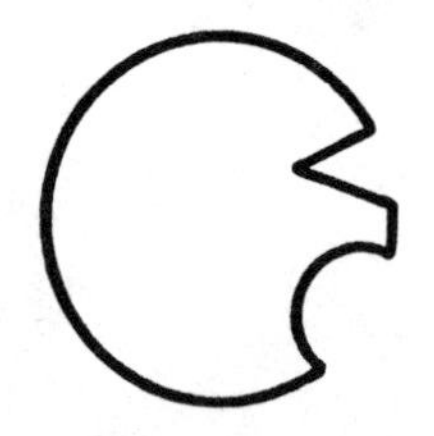

ENZYME AFTER WORKING

MOLECULE SPLIT APART

From *The Chemicals of Life*
by Isaac Asimov
Illustration by John Bradford
Published by New American Library
New York and Scarsborough, Ontario 1954

HOW DO ENZYMES GET THEIR NAMES?

Due to the large number of enzymes, there had to be a system of nomenclature, which was devised by the National Enzyme Commission. We now know that the names of all enzymes end in "-ase." In most cases, the name of the enzyme will also reveal its function; thus, protease is an enzyme that catalyzes (acts upon) proteins; lipase is an enzyme that catalyzes fats; cellulase is an enzyme that acts upon cellulose; and amylase acts upon starches.[3]

There are four categories of food enzymes. They are:

1. Lipase–which serves to break down fat
2. Protease (proteolytic enzymes)–which works to break down protein
3. Cellulase–which assists to break down cellulose
4. Amylase–which breaks down starch

Included in each category are a number of enzymes. For example, trypsin and pepsin act upon proteins, so they are proteolytic enzymes which fall under the category of protease. The enzymes trypsin and pepsin do not end in "-ase" because they were named before the new nomenclature was put into effect. However, even these enzymes have been called "trypsinogenase" and "pepsinogenase."

Do not let the verbiage confuse you or cause you to lose interest in enzymes. The names are not important unless you are going to make a study of enzymes. What is important is that you understand how to get enzymes into your body, their sources, and that without them life cannot exist.

As we become enzyme deficient, we age faster. The more we store up our enzyme reserve, the healthier we will be. You will learn how to increase your enzyme reserve later in this book.

WHERE DO WE GET OUR ENZYMES?

We usually think of enzymes as involved only in digesting our food so we can absorb it. Very seldom do we find informative material that demonstrates that enzymes are involved in every metabolic process. Our immune system, bloodstream, liver, kidneys, spleen, pancreas, as well as our ability to see, think, and breathe depend upon enzymes. The lack of them in any of these areas can prove to be detrimental in degrees of depletion.

Realizing that the lack of enzymes can be a predisposing cause of disease substantiates the importance of enzymes. There is much literature which tries to establish that toxicity and genetics are the predisposing cause of disease. These are true statements, but the important thing to keep in mind is that all cellular activity is initiated by enzymes. Enzymes break down toxic substances so that the body can eliminate them without damaging the eliminative organs.

It is important that we preserve the body's enzyme level at all expenses. There are two ways to preserve and replenish our enzyme level: by eating raw food and by taking enzyme supplements. Supplementing enzymes to our diet and eating a raw food diet are subjects that will be discussed throughout this book. What is important at this time is to understand what happens when we cook our food.

The difference between live (raw) and dead food is enzymatic activity. If you had two seeds and boiled one, which one would grow when placed in soil? There is no question that the unboiled seed would sprout because it has its enzymes intact. All foods provided by nature have an abundance of enzymes when in their raw state.

One characteristic of enzymes is their inability to withstand hot temperatures such as those used in cooking. Consequently, the enzymes are completely destroyed in all foods that are canned, pasteurized, baked, roasted, stewed, or fried.

At 129°F, all enzymes are destroyed. Baking bread kills enzymes. Most butters have no enzymes because they are pasteurized. Canned juices may have vitamins and minerals, but the heating process has killed the enzymes. The roasted breakfast cereals that we feed our children are devoid of enzymes.

When live food substances come in contact with heated water, the enzymes are destroyed quite rapidly.

Dr. Howell, in his book about food enzymes, states, "Enzymes are more or less completely destroyed when heated in water in the temperature range between 48° to 65°C. Long heating at 48°C or short heating at 65°C, kills enzymes. Heating at 60° to 80°C for one-half hour completely kills any enzymes."[4]

Food processing, refining, cooking, and, more recently, the advent of micro-waving are detrimental processes that cause dramatic changes in the food we eat. They have rendered our foods enzyme-deficient, causing imbalances in our organs, acting as a predisposing cause of disease.

Enzymes are always a part of animal and plant life. They are a component part of living matter. Animals in the wild consume large amounts of enzymes as a result of their primary raw food diets. This aids in the digestive process, taking stress off organs such as the pancreas, liver, and spleen which would otherwise have to produce large amounts of enzymes. This causes unwarranted stress on these organs and body tissues, thus decreasing the longevity of the body.

There are three major classes of enzymes: (1) metabolic enzymes (enzymes which work in blood, tissues, and organs), (2) food enzymes from raw food, and (3) digestive enzymes.

Our organs are run by metabolic enzymes. These enzymes take food substances and build them into healthy tissue and have numerous other duties. One authority found ninety-eight enzymes working in the arteries alone.

Since 1968, thirteen hundred enzymes have been identified. A shortage of these enzymes may cause serious health problems.

Nature has placed enzymes in food to aid in the digestive process instead of forcing the enzymes secreted in our bodies to do all of the work. It is to be remembered that we inherited an enzyme reserve at birth and this quantity can be decreased as we age by eating an enzyme-deficient diet.

By eating most of our food cooked, our digestive

systems have to produce all of the enzymes, thus causing an enlargement of the digestive organs. To supply such enzymes, the body draws on its reserve from all organs and tissues, causing a metabolic deficit.

If each of us would take in more exogenous enzymes (those enzymes taken from outside sources), our enzyme reserve would not be depleted at such a rapid pace. This would keep our metabolic enzymes more evenly distributed throughout the organism. This is one of the most health-promoting measures that one could implement into his or her daily lifestyle.

The pancreas, which is of primary importance to our digestive systems, secretes lipase (fat digestive enzymes), amylase (starch digestive enzymes), and protease (protein digestive enzymes). The question is: Does this organ produce all of these enzymes from within itself?

At the University of Toronto, the pancreases from sixteen dogs were removed. These dogs were kept alive for months by the use of insulin and were shown to have normal levels of amylase in their blood streams.[5]

The School of Medicine at Yale University, doing a similar study, found that six dogs had a rise in blood amylase amounting to twenty times the normal amount after the ligation (tying off) of the pancreatic duct.

In a 1936 *Journal of Enzymology,* Fiessinger and Associates could not find a significant change in the lipase content of the blood of dogs fifteen days following the removal of their pancreases. The enzymes that maintain blood levels after the pancreas is removed come from other tissues and organs.[7,8]

It is foolish to believe that an organ weighing just a few ounces (the average pancreas weighs only 85 grams) can supply the enormous amount of enzymes needed in the body day after day, year after year. It can be substantiated that the pancreas receives constituents (enzymes) from the blood and tissues.

Dr. Willstatter has demonstrated the presence of amylase in white blood cells (leukocytes) and also shown that the white blood cells have proteolytic enzymes similar to the secretions of the pancreas.[9] White blood cells have a greater variety of enzymes than does the pancreas. Leukocytes travel through the body, destroying foreign substances in the bloodstream, working as part of the immune system that protects us from disease.

We lose enzymes daily through our sweat, urine, feces, and through all digestive fluids, including salivary and intestinal secretions. Logic dictates that a pancreas weighing only a few ounces could not possibly be responsible for the production of all of the enzymes found throughout the body of a 150-pound man. Enzymes are found in every tissue of the body. Even when the pancreas is removed, the body still maintains a certain enzyme level.

Another point of paramount importance is that a percentage of enzymes which are taken orally, or the ones already present in raw food, can be absorbed in the intestines and utilized in the body's metabolic processes to help prevent enzyme depletion. Before we look at the absorption of enzymes, let's look at other substances that are absorbed through the intestines.

At the University of Illinois, compressed yeast fed to dogs produced positive yeast cultures in the liver, lymph glands, lungs, spleen, and kidneys, proving that whole yeast cells were absorbed.[10]

Dr. Oelgoetz documented in the *American Journal of Digestion* that if patients with low levels of blood amylase are given an extract of whole pancreas which contains amylase, the normal blood level can be restored within one hour and remain normal for days after administration.[11]

It is a well-known fact that unassimilated proteins, yeast cells, carbohydrates, and fats can be reabsorbed into the blood stream, causing allergies, skin diseases, and other illnesses.[12]

Dr. Oelgoetz proved that when pancreatic enzymes are administered to patients having allergies accompanied by low blood levels of enzymes, these levels return to normal and the allergy subsides.

Functional digestive disturbances, hyperacidity, and skin problems were relieved the same way. Clinicians have effectively relieved a variety of skin diseases caused by incompletely digested food materials. The blood provides the ideal environment for enzymes to cause a partial breakdown of undigested materials.

The oral administration of proteolytic enzymes in the treatment of inflammation and sports injuries have been used for years by Max Wolf, M.D. and Karl Ransberger, Ph.D.

In one experiment described in their book, *Enzymes Therapy,* certain enzymes were tagged with a radioactive dye to see if these enzymes could be followed through the digestive tract into the bloodstream. It was shown through electrophoretic investigations that the radioactive dye tagged to the enzymes could be found in the liver, spleen, kidneys, heart, lungs, duodenum, and urine.[13]

If you would like more information showing how enzymes, taken orally, are absorbed and used throughout the body, thus taking the stress off other enzyme-producing organs, see the references in the back of this book.

For now, let's take a look at how the enzymes function in raw food and how they may aid in the digestive process, preparing bodily functions.

HOW FOOD ENZYMES AID DIGESTION

A human being is not maintained by his food intake, but rather, by what is digested. Every food must be broken down by enzymes to simpler building blocks.

Enzymes may be divided into two groups: exogenous (found in raw food), and endogenous (produced within our bodies). The more one gets of the exogenous enzymes, the less will have to be borrowed from other metabolic processes and supplied by the pancreas. The enzymes contained in raw food actually aid in the digestion of that same food when it is chewed. One can live for many years on a cooked food diet but eventually this kind of diet will cause cellular enzyme exhaustion, which lays the foundation for a weak immune system and, ultimately, disease.

Dr. Howell states, "Researchers show that cooked food with the fiber broken down passes through the digestive system more slowly than raw foods. Partially it ferments, rots, putrefies, throwing back into the body toxins, gas, and causing heartburn and degenerative diseases."

It is important to realize that the enzymes in raw food actually digest 5 to 75 percent of the food itself *without* the help of the enzymes secreted by the body. This is called "energy conservation," since the body does not have to supply all of the enzymes needed to digest the food.

At the Institute of Animal Physiology, Agriculture College, Berlin, comparative experiments were done to show that the enzymes contained in raw food aid digestion. Fowl do not have amylase (starch digestive enzyme) in their salivary secretions. In one study, chickens were fed ground barley, which has a good amount of pure starch in it. After five hours, the stomach contents of the fowl were analyzed, showing that eight percent of the starch was digested.[14]

It was also demonstrated by Dr. Boas, as reported by Dr. Howell, that "the enzymes in bananas were activated in the intestines to aid in the digestive process."[15] This also shows that not all enzymes are destroyed in the stomach, but merely inactivated there and then reactivated again in the intestines.

This important point was also proven by a Russian researcher, Dr. Matveev. He demonstrated that oxidase and catalase, which are enzymes supplied from carrot juice, were inactivated in the stomach because of the acidity, and then reactivated again in the alkalinity of the small intestines.[16]

In summary, the enzymes in raw food digested a small percentage of the food. It appeared then that all enzymes are not destroyed in the stomach as once believed, but are actually reactivated again to aid the pancreas in digestion in the small intestine.

We all suffer the consequences of cooked food diets. As Dr. Howell demonstrated, the pancreases of animals that subsist exclusively on raw plant food are much smaller relative to their body weight than humans'.[17]

The pancreas of a human weighing 140 pounds, weighs 85-90 grams; the pancreas of a sheep weighing 85 pounds, weighs only 18.8 grams; the pancreas of a 1005 pound cow weighs only 308 grams; and the pancreas of a horse weighing 1200 pounds, weighs only 330 grams. Calculated as a percentage of body weight, the following figures are presented:

	Body Weight Grams	Pancreas Weight % of Body Weight
Sheep	38,505	0.0490
Cattle	455,265	0.0680
Horse	543,600	0.0603
Man	63,420	0.1400

Notice how little man weighs compared to cattle and horses, and again in comparison, how much larger his pancreas is. This is because the human pancreas must enlarge because it is overworked by consuming a cooked food diet devoid of enzymes.

The fascinating point here is that in humans, the saliva contains amylase to aid the pancreas in starch digestion. Herbivores (animals other than man listed in the chart above) have relatively no amylase in their saliva, and their pancreases still remain their normal size without enlarging.

The answer seems to be that the raw food of herbivores supplies active enzymes which participate in digestion, thus taking stress off the digestive organs, the pancreas, and, in fact, the whole body's metabolism.

Another similar experiment was reported in the

Philippine Journal of Science (Vol. 52) as performed by the School of Hygiene of Public Health in 1933. The school performed 768 postmortem examinations of Filipinos. The observations showed that the pancreases of the Filipinos were twenty-five to fifty percent heavier than Europeans and Americans, because cooked rice was their staple food and was eaten as many as three times a day. This bulk caused their pancreases to be overworked, secreting large amounts of enzymes (particularly amylase), causing an enlargement of the organ. An enlarged organ is often a pathological condition, showing the beginning signs of degeneration.

It is a fact that a cooked food diet causes a larger outpouring of enzymes from our digestive organs. At first thought, it might be presumed that hypertrophied (enlarged) organs are a desirable accommodation, but it has always been shown that enlargement of an organ is accompanied by the excessive function of that organ, followed by exhaustion and degeneration.

Since the enzymes in raw food actually help digest the food in which they are contained and can be absorbed into the blood and used in other metabolic processes, we can assume that taking enzymes or eating a large percentage of raw food will help take the stress off not only the pancreas, but the entire body.

ENZYMES AND LONGEVITY

The comparative study of the enzyme content of the blood, urine, and digestive fluids of the human population can create some very important data. For example, the average diet is predominately heat-treated and possesses only a fraction of its original enzyme content.

It has been shown that young adults have a high value of enzyme reserve in their tissues. In older persons, the potential enzyme tissue reserve is much lower and essentially depleted.

When a young person eats cooked food, there is a greater outpouring of enzymes from the organs and body fluids than in adults. This is because years of eating a cooked food diet has depleted the adult, whereas the young adult's tissue reserve is still at maximum.

A further experiment in relation to saliva and its amylase content was performed at the Michael Reese Hospital in Chicago. One group of young adults between the ages of 21 to 31 and another group of older adults ranging from age 69 to 100 were used in the experiment. It was shown that the younger group had 30 times more amylase in their salivas than the older group.

The increased amount of enzymes is why younger

persons can tolerate a diet of white bread, starches, and predominately cooked food.

However, as our enzyme reserves are depleted over the years, these same foods can cause illnesses such as constipation, blood diseases, bleeding ulcers, bloating, and arthritis. In older individuals, the enzyme content of the body has been depleted and these kinds of foods are not properly digested. They ferment in the digestive tract, producing toxins that are then absorbed into the blood and deposited in the joints and other soft-tissue areas.

A "chronic disease" is a disease that has lingered in the body for many weeks, months, or sometimes years. It has been a constant drag on the body, depleting it of its enzymes, vitamins, minerals, and trace minerals. During chronic disease processes, there is usually a low body reserve of enzymes.

In 111 Japanese patients who had tuberculosis, 82% had lower enzyme contents than normal individuals. As the disease worsened, the enzyme level decreased.[18]

Dr. Volodin, in the *Archives Vendanugskrankh,* found that after studying the enzyme levels in urine, blood, and intestines, the levels were usually decreased in people with diabetes. In many cases, studies of feces showed incomplete digestion of meat and fats.

In five of six diabetic patients, the lipase and trypsin (proteolytic enzymes) of the pancreatic juice were found to be decreased.[19]

Dr. Ottenstein in a similar study pattern showed low blood amylase levels in skin afflictions such as psoriasis, dermatitis, and pruritis.[20]

Another interesting experiment showed that 40 patients suffering from liver diseases such as cirrhosis, hepatitis, and cholecystitis (inflammation of the gallbladder), showed low levels of amylase. It was found that when there was a rise in the blood amylase level, there was an improvement in the general condition of each patient, as well as an improvement in the liver condition.[21]

It is an undisputable fact that during chronic disease we find a lower enzyme content in blood, urine, feces, and tissues. In acute diseases, and sometimes at the beginning of chronic diseases, the enzyme content is often found to be high. This shows that the body has a reserve and the tissues are not yet depleted; consequently, there is a larger outpouring of enzymes in the battle against disease. As the disease progresses, the body's enzyme content is lowered.

This correlation between a diminished enzyme content during chronic disease and old age is often misunderstood. A low enzyme content in old age is often looked upon as "normal." A low content during chronic diseases is considered a pathological state.

The truth of the matter is that age is not so much a matter of chronology, but rather is a matter of the integrity of the body tissues. These tissues depend upon the amount of enzymes present to carry on the metabolism of every cell in the body. It is common to find a 60 year-old man or woman with a body of someone in his or her 40s.

There is a definite correlation between the amount of enzymes an individual possesses and the amount of energy they have. Increasing age causes a slow decrease in enzyme reserve. When the enzyme level becomes so low that metabolism suffers, death will finally result.

Any time the metabolism is falsely stimulated by coffee, a high protein diet, or other stimulants, the metabolism increases, enzymes are used up, a false energy output is experienced, and the individual feels a sense of well being. However, the end result will be *lower* energy, a more rapid burnout of enzymes, and premature old age.

At Brown University, a group of 158 animals were overfed. They lived, on the average, 29.6 days. Another group was maintained on a starvation diet, given only small amounts of food and fluid. They lived, on the average, 39.19 days—an increase of about 40%.[22] At the very least, this study should make each of us look at our own intake and determine if we are indeed over-ingesting.

A high protein diet is very stimulating to the body but can cause serious damage. When the diet consists of more protein than is needed, the excess is broken down by enzymes in the liver and kidneys. The major by-product of protein breakdown is urea, which is a diuretic. Urea stimulates the kidneys to produce more urine. Along with water, minerals are lost in the urine. One of the most important minerals lost is calcium.

Experiments have shown that when subjects consumed 75 grams of protein daily, even with an intake as high as 1400 milligrams of calcium, more calcium was lost in the urine than was actually absorbed.[23]

This deficiency must be made up by the body's calcium reserve, which is taken from the bones. Deficient bones are a stepping stone to osteoporosis (a condition that causes bones to break easily).

The aforementioned experiments all have shown that when excessive amounts of protein, or food in general, are eaten, there is a corresponding decrease in enzyme, vitamin, and mineral levels.

At the University of Toronto, a team of scientists showed that life runs its course in direct proportion to the "catabolic rate." (The catabolic rate is a measure of the rapidity of the wear and tear of the body or the rate of tissue breakdown. This is in direct proportion to the aging process.) This tissue breakdown is performed by enzymes. The faster the breakdown, the more enzymes are used up.

Our enzyme reserve can be used up rapidly, or it can be preserved. Taking enzyme supplements and eating raw foods are ways to add enzymes to our enzyme reserve and add to our energy level.

Dr. Howell further states that, "Enzymes are a true yardstick of vitality. Enzymes offer an important means of calculating the vital energy of an organism. That which we call energy, vital force, nerve energy, and strength, may be synonymous with enzyme activity."

Our logic tells us that the buildup and the breakdown of tissues is performed by enzymes. In other words, our metabolism is maintained by enzyme activity. When our enzyme level is lowered, our metabolism is lowered, and so is our energy level.

Do not misunderstand this statement. We are not

saying that the source of life is enzymes, but that there is a correlation between enzyme levels and the youth of the tissues of an organism and its energy levels.

Investigations have shown that in warm temperatures, enzymes are used up more rapidly than in cool temperatures. When starch-digesting enzymes are added to potato starch and placed in a room that has a temperature of 80°F, the starch is digested much more rapidly than starch with enzymes placed in a temperature of 40°F. As the temperature increases, the enzymes work harder and are used up faster.

The prevalent thought here is that enzymes are not actually used up, but many tests have shown that various enzymes are found in the urine after fevers and athletic activity. Enzymes are found in the urine, feces, and sweat, along with the used-up substances from proteins, enzymes, fats, carbohydrates, vitamins, and minerals.

Other food substances, such as vitamins and minerals, are replaced daily in our food intake. Not enough attention is placed on taking enzyme supplements or eating raw food. If we do not replenish our enzyme level and only consider vitamins and minerals, we defeat ourselves. The body must replace enzymes from within itself, stealing enzymes from all parts of the body—which in the end causes exhaustion, premature aging, and a low energy system.

The utilization of vitamins depends upon enzymes, and enzymes often depend on vitamins. Under clinical observation, it has been shown that when taking vitamins combined in capsules with enzymes, smaller amounts of vitamins and minerals are needed.

A good example of this phenomenon was seen in a patient (known personally by the author) who needed 70 milligrams of zinc daily to overcome a severe depletion. When zinc was combined with certain enzymes, this patient needed only 3 micrograms daily—a drastic reduction.

It seems that the body needs smaller amounts of vitamins and minerals when combined with enzymes. As good consumers, all of us are interested in saving money. This is possible by reducing our intakes of vitamins and minerals to maintain our daily requirements. It has been the experience of many clinicians that a patient is more likely to follow a health regimen if it can be made as easy and as practical as possible.

ENZYMES AND THEIR RELATIONSHIP TO DISEASE

Enzymes are a part of every metabolic process in the body—from the working of our glands to the proper functioning of our immune system. This is a general way of looking at the working of enzymes in our bodies. Specifically, from the standpoint of individual disease, enzymes can be related to every known disease.

When certain enzymes are in excess in the bloodstream, they can be used for diagnosing specific diseases. Lipase breaks down triglycerides into fatty acids and glycerol. The pancreas secretes a large amount of lipase (it draws much of the lipase from the blood) into the digestive tract. Where lipase is found in elevated levels in the blood, it can point specifically to a pancreatic disease. Serum lipase levels rise when pancreatic inflammation occurs.

Acid phosphatase breaks down phosphates in the blood. This enzyme is found in the prostate gland, red blood cells, and platelets. In these locations, blood serum enzyme levels are measured primarily to evaluate the presence of prostatic carcinoma (a malignant growth). Enzymes not only play a role in digestion, but also during disease and other metabolic processes.

The speed of the metabolism is determined by the

activity of enzymes. The more rapidly the metabolic process, the more enzymes are required to participate, and thus the faster the enzymes are used up.

During exercise and in acute diseases, the enzyme level can be found to be increased. Dr. Gerner, in a study done in 1933, made 300 amylase level determinations on 115 subjects, representing 28 different acute infectious diseases. The urinary amylase was increased in 73% of them. During pneumonia, acute appendicitis, malaria, pulmonary tuberculosis, fevers of all types, and children's diseases, enzymes were found to be elevated in blood, urine, and feces.[24]

Any increase in metabolic activity, whether it is associated with fevers, heart action (exercise), digestion, muscular work, or pregnancy, is paralleled by an increase in enzyme activity.

It is important to note that enzyme activity increases as temperatures increase and is present in most acute disease conditions, fevers, and during exercise. In other words, enzymes perform more work during fevers of 104°F than at normal body temperatures.

It is evident that if enzymes respond to fevers and infections, they have a direct relationship to the defense mechanism in our bodies. Keep in mind, however, that similarly when fevers decrease, so does the enzyme activity.

This understanding that enzymes are used up more rapidly during disease, detoxification, digestion (in fact, every time our metabolism is speeded up) shows the importance of supporting our enzyme reserves at all times. It is true that during youth, the body can respond to such stimulation, but only to borrow against future enzyme resources.

There is a connection between the strength of our immune systems and our enzyme levels. The greater the amount of enzyme reserves, the stronger our immune systems, the healthier and stronger we will be. It has been clearly stated that enzyme activity increases during digestion and also during any other increase in metabolism, such as that found in acute diseases. But what is the exact correlation between our immune system and enzymes?

Our white blood cells (leukocytes) are responsible for destroying foreign, disease-producing substances in the blood and lymph fluids in the body. During acute diseases and infections, the white blood cell count increases to help fight off these pathologies.

Dr. Willstatter, in an early enzyme research study, demonstrated that there were 8 different amylase enzymes found in leukocytes.[25] Investigations also have shown that leukocytes contain proteolytic and lipolytic enzymes which are also common to those secreted by the pancreas. These enzymes act very much like the enzymes that are in our digestive tract (breaking down proteins, fats, and carbohydrates that have been absorbed by the blood, causing disease conditions.)[26, 27]

Enzymes act as scavengers in the body. They latch onto foreign substances and reduce them to a disposable form. They also prevent the arteries from clogging up and joints from becoming gummed up.

It was always thought that the pancreas produced all of these enzymes, but as was mentioned earlier, this idea is erroneous. As small as the pancreas is, it couldn't possibly produce all of the enzymes found in the muscles, glands, and tissues, as well as to produce those used up daily in digestion and those lost in the sweat, urine, and feces. Enzymes are produced by all the tissues and cells of the body. And, in fact, it has been shown that the enzymes found in the white blood cells act very much like the enzymes found in the pancreas, especially the proteolytic enzymes.

Dr. Willstatter found it remarkable how closely the enzyme systems of white blood cells and the pancreatic glands agree with one another. Since the same enzymes are found in the white blood cells as are found in the pancreas, and since white blood cells transport these enzymes throughout the body, it seems that the pancreas and other enzyme-secreting glands receive a great portion of these enzymes via the leukocytes.

After eating a cooked food meal, when digestive enzymes are desperately needed, the white blood cell count increases, seemingly to aid in the digestive process. Since every metabolic process is, at all times, interdependent and interrelated, this increase in the white blood cell count after the ingestion of a cooked meal, indicates a definite compensatory measure.

The body must supply a large amount of digestive enzymes because the enzymes that were once present in the food, were destroyed by the heating process. Dr. Kautchakoff, in his book that demonstrates the relationship of cooking and its effects on our systems, showed that there was an increase in white blood cells after eating a cooked food meal.[28]

This increase in leukocytes is needed to transport enzymes to the digestive tract. Kautchakoff also demonstrated that after a raw food meal, there was no substantial increase in leukocytes, showing that the body has to work much harder to produce and transport enzymes for digestion after a cooked food meal.

It is important to remember that enzymes in raw food aid in the digestive process and that their action removes the stress of having to borrow them from the body's enzyme reserve, particularly from the white blood cell count (an important part of our immune systems).

A most important point in Kautchakoff's experiment is that "leukocytosis" (increased white blood cell count) is a term which describes a medical pathology. Anytime the white blood cell count is increased to any great extent, it is considered that an acute illness or infection is present somewhere in the body. During acute diseases, enzyme levels rise. During chronic diseases, the body enzyme levels are decreased. The pancreas and digestive tract are weakened, for example, during diabetes, cancer, or chronic intestinal problems.

During the course of a chronic disease, the immune system also shows signs of great expenditure. The correlation is clear. Enzymes are found to be related to all diseases via the immune system, whether the disease is acute or chronic. Our enzyme levels must be maintained at any expense to help maintain vitality, endurance, and to prevent disease.

If the pancreatic output of enzymes is hindered, the whole body is affected. If a disease is present, enzymes are used up to fight the condition and the pancreas is affected. You can see how eating mostly cooked food all our lives, or trying to overcome a chronic disease while still eating this type of diet, can be detrimental.

ENDOCRINE SYSTEM AND ENZYMES

A cooked food diet not only kills the enzymes in food but causes the endocrine glands to become overworked and encourages the development of diseases such as hypoglycemia and obesity.

Columbia University published a paper entitled, "Comparative Experiments with Canned, Home Cooked and Raw Foods." Canned food, because of being cooked and/or preserved, is completely devoid of enzymes. It exerts a powerful stimulating effect on the endocrine glands, which can cause an increase in body weight.

It has often been said that cooking makes food easier to digest and assimilate. This may be true, but if cooked food raises the white blood cell count, it can cause weight gain, and also robs the body of enzymes which otherwise would have been used to maintain metabolism. Therefore in the long run, it appears that it cannot be the best diet regimen to follow.

The endocrine system and the nervous system cooperate to regulate the appetite. The glands know when the body has had enough food and will shut off the food craving.

Eating mostly raw food takes the stress off the endocrine system. Sugar and processed foods disrupt the endocrine balance because of their high caloric content. If the glands know the organism has had enough calories, but if the nutrients and enzymes that usually accompany food aren't present because of overcooking, the glands, not finding these nutrients, overstimulate the digestive organs, demanding more food than is needed to maintain strength and vitality. This results in the oversecretion of hormones, overeating, obesity, and, finally, exhaustion of the hormone-producing glands, not to mention the enzyme reserve it depletes trying to carry on the increased metabolic activity.

The false feeling of well-being is caused by the overstimulation of the pituitary gland. This gland is considered to be the "master gland" because it sends hormones to all of the other glands such as the thyroid, adrenals, reproductive glands, and pancreas.

One can see the overshadowing damage that is done by eating enzyme-free, overcooked food. Large amounts of enzymes are used up in eating overcooked food, leaving the organs and tissues without their rightful share.

By eating properly, the breakdown and malfunction of organs not directly involved in digestion are hindered. Most diagnosticians are unaware of this fact. Since disease rarely involves just one or two organs, usually the entire body is affected. Enzymes, the circulatory system, nervous system, endocrine system, and the digestive system are connected and interdependent in life processes.

OBESITY AND CIRCULATORY DISEASES

Enzymes are not only important in terms of digestion, but they also have other metabolic functions. They are found in tissues throughout the body, and when they are deficient in certain areas, they can be a major factor in the cause of obesity and circulatory diseases. By observing enzyme therapy in different countries, it has been found that positive results have been obtained by administering enzymes during certain diseases.

Raw animal tissues and plants have sizeable amounts of lipase if unheated. It has been reported that in human obesity, the lipase content of the fatty tissue is decreased.

Dr. David Galton of Tufts University School of Medicine examined eleven individuals weighing from 230-240 lbs. and found enzyme deficiencies in their fatty tissue. Not only were the fatty tissues lacking in lipase, but fatty tumors (lipomas) were also found to be deficient in lipase.[29]

Lipase is the enzyme that aids the body in fat storage and the breaking down of fats. Animals in hibernation lose pounds of weight while sleeping because of lipase activity. Without lipase, fat stagnates and accumulates in the arteries, capillaries, and other organs.

Why is there a deficiency of lipase in fatty tissues and in obese individuals? When food is cooked, the lipase, which aids in fat digestion, the burning of fat for energy, and the storage and distribution of fat, is absent. Many experiments have been done make this determination. The problem is that these experiments were done using cooked food instead of raw food. As a result, the experiments demonstrate only the fact that cooked calories put on weight, not necessarily raw food calories.

Foods high in calories, such as meat and potatoes, have high enzyme values when raw. When cooked, these same foods provide few, if any, enzymes.

Of course, many foods are not eaten in their raw state, although some Eskimos do eat their meat raw. After cooking, these foods have to be used by the body via enzymes. These enzymes are extracted from the body's own tissues. If there is an overabundance of cooked calories, they are stored in the body tissue as fat.

This fat accumulates in the liver, kidneys, arteries, and capillaries. The stress on the body caused by enzyme-free food not only causes an increase in body weight, but also causes the internal organs to change.

For example, a heat-treated, refined food diet causes drastic changes in the size and appearance of the pituitary gland. This relationship between enzymes and our glands was shown when the surgical removal of the glands in animals led to changes in the enzyme level of the blood.[30] Enzymes affect hormone-producing glands, and hormones influence enzyme levels.

The glandular secretions of the pancreas and pituitary glands become exhausted from overstimulation resulting from a cooked food diet. The body becomes sluggish, thyroid function also becomes exhausted, and the individual gains weight.

Raw calories are relatively non-stimulating to glands and help to stabilize body weight. Farmers have proven this fact by feeding raw potatoes to hogs. On this diet, they did not get fat. Cooking the potatoes, however, produced rapid weight gain in the hogs, and they could be sold for more money.

Think of the meat on the grocery shelves from animals like these with high fat content and low enzyme levels. Once cooked, the meat has fewer enzymes. The fat is saturated, which makes it difficult to digest. And finally, the fat accumulates in our arteries.

Dr. G.E. Burch of Tulane University demonstrated some interesting facts about the cause of obesity. He showed that young, overfed animals develop more fat cells than underfed ones.[31]

Infants who are overfed can develop three times as many fat cells as is normal. When a normal person gains weight, he or she may get what is called "pleasingly plump," but when a person who has been overfed as an infant and who has accumulated many more fat cells that are filled to excess, obesity results.

Both types of people can eat the same amount of food, but the one who has more fat cells and, consequently, more room to store fat, puts on weight much more easily. A good way to help these people is to add raw foods to their diet with the addition of enzyme supplements. Personally, I lost 70 lbs. doing exactly this, and by limiting my diet to mostly raw foods, I haven't gained a pound in twelve years.

If one wants to reduce and to keep his or her weight down, eating fewer meals per day will be beneficial. Frequent eating and snacking can decrease the enzyme level of the body and cause weight gain.

In an experiment carried out at two different universities, the same results were found in two test groups. Two groups of rats were used. One group was fed every two hours; the other, once a day. The rats that ate once a day lived 17% longer, had lower body weight, and higher enzyme activity in their pancreas and fat cells.[32]

It is important to remember that aging corresponds to diminishing enzyme levels. In this same experiment, it was found that enzyme activities in the tissues became weaker as the rats became older.

Enzyme deficiencies can also be the cause of circulatory diseases, high blood pressure, and other blood vessel problems. In particular, the enzyme lipase is involved in these diseases. Remember, lipase is the enzyme that digests, dissolves, and helps break down fats throughout the body.

It has already been shown that younger people have a higher enzyme content in their blood and tissues than older folks. Doctors Berker and Meyers found deficient lipase in individuals who had arteriosclerosis, high blood pressure, and slow fat absorption.

Their investigations also showed that the blood of individuals 77 years of age had only half the amount of blood lipase than a group of 27 year olds.

Another observation made was that fat, not digested properly by lipase, can be absorbed in an undigested, adulterated state. This fat is later found in the blood vessels and arteries, causing conditions such as arteriosclerosis (hardening of the arteries), high blood pressure, and high cholesterol.

These fatty deposits clog the blood vessels and block blood from getting to the heart, which can result in heart attacks. The heart must work harder to push the blood through congested vessels—causing high blood pressure and an enlarged heart. The offenders in these problems are saturated fats (particularly from animal products), hydrogenated fats, and polyunsaturated fats.

We have all been told that polyunsaturated fats lower cholesterol levels. This is true, but they work like a drug. In a fine book called *The McDougall Plan,* it is stated that "...polyunsaturated vegetable oils can be a health hazard. When consumed, they act like a cholesterol-lowering drug. They drive large amounts of stored cholesterol from the body tissues through the liver to the gall bladder and into the colon. In the bowel, the excreted cholesterol may be involved in the cause of cancer of the colon."[33]

Rats fed on diets high in cholesterol and polyunsaturated oils have more colon cancer than rats fed cholesterol and saturated fats. Although experiments with rats and other animals can't always be correlated to human conditions, it is a well known fact that elements that cause cancer in animals can also cause cancer in humans.

Fats impair the function of blood cells in the immune system by slowing down cell circulation. This may be the reason why obese individuals seem to be prone to infections. High levels of fat in the blood also block the action of insulin, which aids in the tissue absorption of sugar. This allows sugar levels to rise in the blood, which can be a contributing factor in diabetes.

Why do we find such a high percentage of circulatory diseases in civilized nations, and not in uncivilized groups of people? Primitive Eskimos, who consume as much as ten pounds of meat daily, consisting of raw fish and blubber, have few signs of circulatory disease. They don't develop vascular diseases because most of their food is raw. The enzymes, in particular, lipase, are still active in raw meat, which helps to prevent circulatory diseases.

At Stanford University, investigators showed that patients who had hardening of the arteries were deficient in lipase. The more advanced the disease, the more enzyme deficient the patient.

In another study, three British physicians demonstrated that in normal individuals, the enzyme lipase was found in blood serum. But patients who had hardening of the arteries were found deficient in blood enzymes. When lipase was given to patients with slow fat metabolism and blood fat problems, there was an immediate improvement in fat metabolism.[34]

In wild animals and in primitive Eskimos who consume raw meat, enzymes keep blood vessels clear of fatty deposits by promoting proper digestion and breakdown of fats, both in the digestive tract and in the liver. Dr. Maynard Murray, a research scientist, dissected more than 3,000 whales that had a three to six inch layer of fat around their bodies. These whales had no signs of circulatory diseases. Their diet is raw fish. Isolated Eskimo tribes live on raw fish and meat.

In 1926, Dr. William A. Thomas studied primitive Eskimos and found no signs of kidney or vascular diseases. In the adult Eskimos, ranging in age from 40 to 60, the average blood pressure reading was 129/76.

It should be remembered that these people were primitive. More civilized Eskimos who settled in the Hudson Bay area, close to trading posts, used cooked food and white flour products. They gave up their primitive diet, and consequently their good health. One can now find arteriosclerosis and high blood pressure among these people. The only factor that is found to be different among primitive and civilized Eskimos is their diets.

Primitive Eskimos predigest their food so that when they eat, their bodies don't have to do all of the digestive work, thereby depleting their systems. They catch fish and bury it until it partially decays. All live tissue contains enzymes and these enzymes break down and predigest the food. The fish is then unearthed and eaten. It is called "high fish" because it gives them strength and endurance. The body doesn't have to secrete large amounts of enzymes because the food is in this easily digestible state. Digestive energy is conserved, leaving more for work, metabolic functions, thinking, and exercise.

The Eskimos know that when they work their dogs for weeks, the animals become worn and tired. When fed fresh, raw fish, the dogs become weak and thin, but when fed "high fish," they gain weight and strength.

This "predigestion" concept is not new. Asiatic peoples have improved soybeans by adding enzymes from fungal plants to them, thus predigesting the beans. As a result, we find foods such as tofu, tempeh, and miso in health food stores in the United States.

Cheeses and meats are aged by enzymes. In Lebanon, a dish of crushed raw lamb and wheat is left to age so that the enzymes predigest the food before it is eaten.

In summary, enzymes are preserved in live, uncooked food. These enzymes not only predigest the food by fermentation processes, but also aid in digesting the food when eaten in its raw state. Within the blood stream and tissues of the body, enzymes act as scavengers, breaking down cholesterol and fatty deposits, and assisting in the overall detoxification process.

We do not say that all food should be eaten raw by everyone, but a great majority of food should be. Studying the works of Viktoras Kulvinskas, author of *Survival into the 21st Century*, one can find information about raw food diets, sprouting, and predigested foods.

RAW FOOD DIET AND PREDIGESTION

It is a tremendous advantage to know how to use raw foods and supplemental enzymes in the diet. Well-known naturalists such as Arnold Ehret, Dr. Ann Wigmore, George J. Drews, and Viktoras Kulvinskas are people who have shown the healing effects of raw foods. It is a fact that the enzymes in raw foods aid in digestion. This enzyme activity takes place in what Dr. Howell has called the "enzyme-stomach," or the upper part of the stomach. Anatomically speaking, it is called the "cardial" and "fundic" parts of the stomach where foods are predigested by enzymes. I will refer to the upper portion of the stomach as the "enzyme-stomach." (See Diagram 2 on the opposite page.)

The enzyme stomach is not original to the human race. Cattle and sheep have no enzymes in their saliva, but have four stomachs. Only one stomach secretes enzymes; the other stomachs allow the enzymes within the food do the partial-digesting or predigesting.

Dolphins and whales have three stomachs. One is devoid of enzymes when the food enzymes do the predigesting. There have been as many as thirty-two seals found in the enzyme-stomach of one whale at one time. Since there is no enzyme secretion in this

DIAGRAM 2

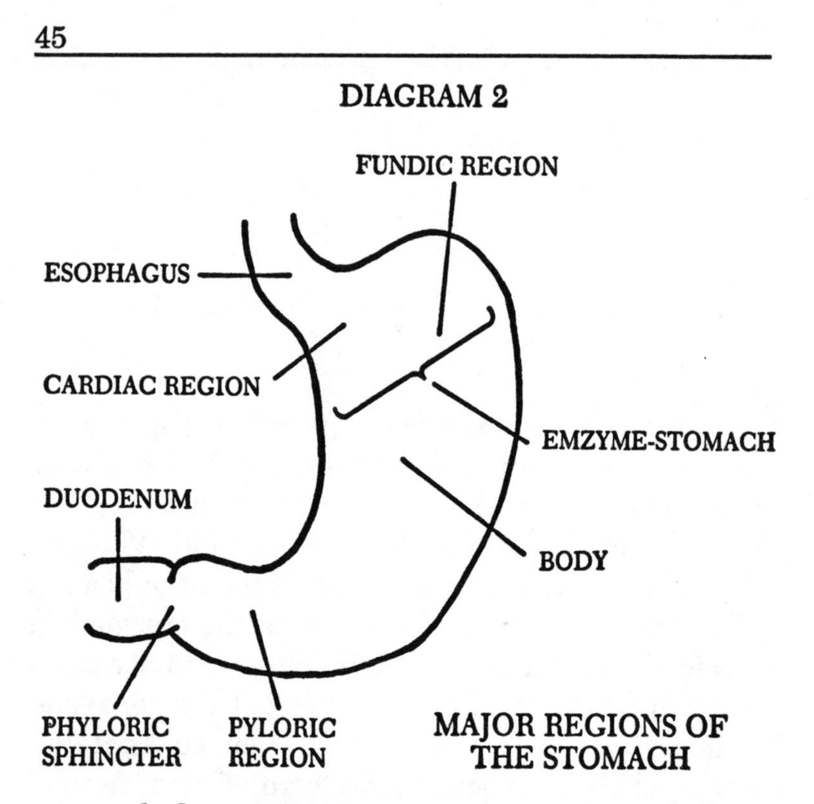

MAJOR REGIONS OF THE STOMACH

stomach, how does all of the flesh get broken down into small enough amounts in order to pass the small opening which connects the enzyme-stomach to the other stomachs? The enzymes contained in the flesh of the eaten animals themselves seem to be the only answer.

After the death of animal bodies, the tissues become acidic. This acid releases an enzyme called "cathepsin" that begins the predigestion of the flesh protein. This enzyme and others predigest the food and remain active for long periods of time. The whale or other animals that have consumed their food benefit tremendously from this predigestion. Their bodies do not have to secrete large amounts of enzymes to digest the food.

We have already shown that enzymes help digest the food in which they are contained if they are not destroyed by heat during cooking. These enzymes not only work in the stomach, but continue their activity in the small intestine. Popular opinion seems to be that the acid in the stomach destroys the enzymes in food. Research has shown this to be untrue.

Northwestern University scientists have shown that supplemental enzymes pass through the stomach unharmed. In one study, amylase that was derived from germinated barley digested starch in the stomach and passed into the small intestine where it continued digestion.

It was also thought that only protein was digested in the stomach. Research by Dr. Beazell, described in *The Journal of Laboratory and Clinical Medicine,* showed that several times more starch than protein was digested in the stomach within the first hour.

Olaf Berglim, Professor of Physiology at the Illinois College of Medicine, has also shown this to be true. He gave his subjects a meal of mashed potatoes and bread. Both foods contain large amounts of starch. The stomach contents were retrieved after 45 minutes. Seventy-six percent of the starch in the mashed potatoes was digested and fifty-nine percent of the bread starch was digested.

Research has shown that for the first forty-five minutes to one hour, a good percentage of food can be predigested in the stomach by food enzymes or supplemental enzymes before reaching the small intestine.

At this time, the pancreas secretes its protein, fat,

and starch-digestive enzymes into the duodenum (the first part of the small intestine).

The pancreas can be put under tremendous stress if the food is not properly predigested at this point. It must draw enzymes from the whole body to secrete the designated amount.

The more digestion that takes place before the food reaches the small intestine, the better for the integrity, strength, and immunity of the whole body. The pancreas is one of the first organs to dysfunction during diabetes and other chronic illnesses.

The importance of this food-enzyme stomach, where predigestion takes place, cannot be overestimated. It is the upper part of the stomach where no acid secretion or peristalsis takes place for one-half to one hour after food is eaten. This is when supplemental enzymes and the natural enzymes contained in food do their predigestive work.

Gray's Anatomy cites the authority Walter B. Cannon, who demonstrated that the human stomach "consists of two parts physiologically distinct." His work states:

"The cardiac portion of the stomach is a food reservoir in which salivary digestion continues; the pyloric portion is the seat of active gastric digestion. There are no peristaltic waves in the cardiac portion."

Predigestion by exogenous (outside) enzymes is widespread in nature. Our enzyme potential has other and more useful and taxing work to do than merely making endogenous digestive enzymes to digest food.[35]

The body is greatly relieved when predigestion

takes place. The draw on metabolic enzymes is kept at a minimum.

There is a law called the "Adaptive Secretion of Digestive Enzymes." This law states that the more digestion that is accomplished by food enzymes or supplemental enzymes in the stomach, the fewer enzymes need to be secreted by the pancreas and intestines, conserving the body's enzymes for metabolic processes such as repairing tissues, organs, and performing other functions.

If food is over-cooked and its enzymes destroyed, the only enzymes that get mixed with the food are the ones contained in saliva. Some starch digestion may take place in the stomach from the saliva amylase. The protein is acted upon by the stomach pepsin, but mostly in the lower part of the stomach. In both instances, no help from outside enzymes is demonstrated.

The fat remains practically untouched only to wait until it moves into the small intestine for the pancreatic secretions of lipase. The food remains in the food enzyme stomach for its allotted time and practically no predigestion takes place, except for the starch.

Cooked foods, especially those high in protein, can begin to putrify. The byproducts of putrification are toxins that are absorbed into the bloodstream and deposited at body sites far removed from the intestinal tract.

At this point, one can see how valuable enzymes are in helping to keep the blood clear of poisons. It has been estimated that 80% of diseases are caused by improperly digested foods and their by-products being absorbed into the body.[36]

A diet containing mostly cooked food has proven to be detrimental in more ways than one.

> Cooking does not improve the nutritional value of food. It destroys or makes unavailable 85% of the original nutrients. Cooked food is totally lacking in enzymes; most of the protein has been destroyed or converted to new forms which are either not digestible by body enzymes or digested with difficulty; many of the vitamins have lost their vitality. To purchase organic food and then to waste precious hours in destroying most of the nutrients is poor economy and unsound ecology. Francis Pottenger, M.D., carried out a 10-year experiment using 900 cats, which were placed on controlled diets. The cats on raw food produced healthy kittens from generation to generation. Those on cooked food developed our modern ailments: heart, kidney and thyroid disease, pneumonia, paralysis, loss of teeth, difficulty in labor, diminished or perverted sexual interest, diarrhea, and irritability. Liver impairment on cooked protein was progressive, the bile in the stool becoming so toxic that even weeds refused to grow in soil fertilized by the cats' excrement. The first generation of kittens were sick and

> abnormal; the second generation were often born dead or diseased; by the third generation, the mother was sterile.[37]

Zoologists know that captured animals fed a human diet develop human diseases such as gastritis, duodenitis, colitis, liver diseases, anemia, thyroid diseases, arthritis, and circulatory problems.

I am not advocating that everyone should exist on a raw food diet. Some individuals don't have the inclination nor the interest to even try. However, improving one's diet by taking supplemental enzymes or eating a good quantity of predigested foods can be beneficial.

Naturally, predigested foods include sprouts and fruits. This means that the proteins, starches, and fats are already predigested and high in enzymes before they are consumed. Sprouting seeds is simple and beneficial to your health. There are several books in health food stores which can explain sprouting procedures.

BODY, MIND, AND ENZYMES

Since it seems that our happiness depends largely upon the thoughts we think, it is impossible to think positively at all times when we are toxic, stiff, and are experiencing low blood sugar levels. Our brain exists on large amounts of sugar and oxygen. When this supply is low, we experience a lack of concentration, insomnia, lethargy, irritability, and confusion.

A lack of enzymes, oxygen, and sugar supplies to the cells of our body can cause hypoglycemia. Hypoglycemia is a disorder resulting from too low blood sugar, and sugar is the fuel for our cells. Authorities have estimated that anywhere from *ten to one-hundred million Americans* suffer from hypoglycemia.

Since hypoglycemia is a malfunction of our fuel supply, every organ is affected. Here's how. As the sugar level drops, the metabolism of every organ drops, resulting in fatigue and psychosomatic problems. The brain is nourished exclusively by glucose and oxygen. A drop in one's blood sugar can cause mental fatigue and depression.

The endocrine glands, especially the pituitary, adrenals, thyroid, and pancreas, control the sugar level. The pancreas secretes insulin. Insulin causes a decrease in blood sugar. Insulin facilitates the movement of glucose (blood sugar) to leave the blood and enter the cells. Insulin also stimulates the liver and muscle cells to convert glucose into glycogen, which is a carbohydrate and the chief storage compound of sugar in the body.

The adrenal glands secrete a hormone called "epinephrine" that causes the stored sugar (glycogen) to break down into glucose, which then enters the blood to raise the blood sugar.

The thyroid gland secretes hormones that control the rate at which the body uses oxygen. Also its hormones increase the rate of energy released from carbohydrates.

All of these glands are controlled by the pituitary gland, which, in turn, is controlled by an area of the brain called the "hypothalamus." The hypothalamus receives information from all parts of the body via the nervous system. This includes (whether hungry or not) a person's emotional state, body temperature, and blood nutrient concentration, among other things.

It has been shown that the pituitary and other organs can enlarge, become exhausted, and be susceptible to disease when a deficiency of enzymes is present. When there is a lack of blood amylase, blood sugar levels can be higher than normal. With the addition of the enzyme amylase, blood sugar levels have been lowered.

In experiments done by Grubler and Myers, they showed that by giving amylase preparations to normal individuals after eating 80 grams of glucose, the blood sugar level was maintained.[38]

Reports show that oral or intravenous injection of

amylase causes a lowering of blood sugar levels in diabetics. Bassler showed that 86% of the diabetics that he examined had a deficiency of amylase in their intestinal secretions. After administering amylase to a majority of these patients, 50% of the diabetics who were users of insulin could control their blood sugar levels without the use of insulin.[39] Amylase seems to help the storage and utilization of sugar in the blood.

Cooked food, in which most of the amylase and other enzymes are destroyed, has a tremendous effect on the blood sugar levels. At George Washington University Hospital, 50 grams of raw starch were fed to hospital patients. The blood sugar showed an average increase of 1 mg. per 100 cc. in one-half hour, a decrease of 1.2 mg. in one hour, and a decrease of 3 mg. in two hours.

When 50 grams of cooked starch was given, the average increase in blood sugar was 56 mg. in one-half hour, then dropped to 51 mg. in one hour, then down to 11 mg. in two hours after eating the meal.

In the diagram on the next page, notice the difference between the blood sugar levels when cooked and when raw starch with the enzymes present, is eaten. The blood sugar in cooked starch rose to 56 mg. in one-half hour opposed to the uncooked starch increase of only 1 mg. After two hours, the cooked food starch eaters' blood sugar level fell to 11 mg., a 45 mg. drop in blood sugar. This resulted in fatigue, anxiety, and the other aforementioned symptoms. The raw starch eaters only experienced a drop of 1 mg. to 3 mg. in two hours. A much more steady metabolic rate and emotional stability was experienced by the raw food eaters.

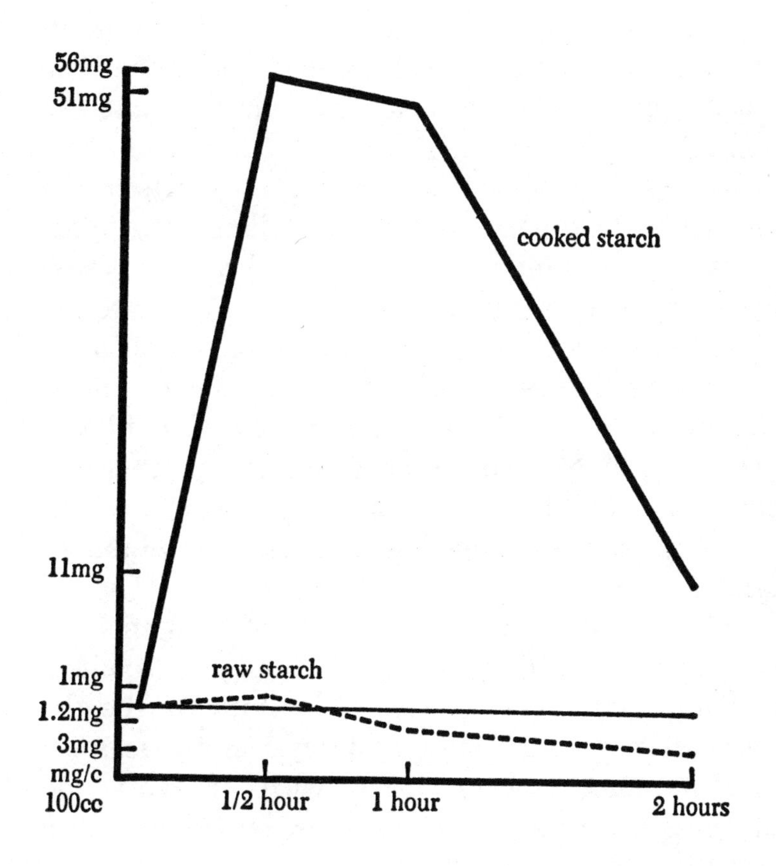

The endocrine glands need trace minerals and vitamins to function properly. An example of this is that of the thyroid gland needing iodine, and the adrenal glands needing vitamin C. Overcooked food is deficient not only in enzymes, but also in nutrients. These deficiencies cause many problems.

The glands of the body are controlled by stimuli from the brain to secrete their hormones. When the blood sugar level drops below normal, the pancreas and adrenal glands are called upon to secrete their hormones. When there is a lack of nutrients in the blood which support the endocrine glands, the hypothalamus stimulates the appetite and causes a craving for food. The more that cooked food is eaten, the more there will be hormone stimulation, resulting in overeating. Excessive eating can cause one to be overweight and obese. Obesity can be the cause of heart problems, high blood pressure, and many other diseases. Quick rising and falling blood sugar levels in the body cause emotional swings and mental imbalances.

Finally, the endocrine glands, deficient in their secretions from trying to keep the body metabolism normal, become exhausted. This state can be the foundation of both mental and physical diseases.

Enzymes have as much to do with our mental and physical health as any other considered element of nutrition. This is a subject that has not been given enough attention in today's literature. I hope this book will enlighten you about the importance of food enzymes in your system as well as bring it to the public's eye.

DETOXIFICATION AND ENZYMES

There are several methods that have been used to detoxify the body: fasting, purging, vegetarian-type diets, macrobiotic diets, the "grape cure," and others. All of these have been successful to some people on their own. But at the same time, none of these regimes will work all the time. The question is: what can be done to improve all of these methods of cleansing the body to make them more proficient?

Anytime cooked food is eaten, it must be digested by enzymes. The waste left over from the digestion of these foods and the toxins contained within the foods themselves are broken down by scavenger enzymes in the body's immune system. The body draws enzymes from all of its tissue resources to do the job, decreasing its enzyme level.

Concentrated foods, such as heavy starches (breads and flour products), animal proteins, and fried foods, are more difficult to break down and consequently more enzymes are needed. This is why, in most situations, a vegetarian diet is used for cleansing purposes. A good percentage of these diets are raw, which add some enzymes to the body and to the food-enzyme stomach. Cooked vegetables and grains use some of the body's enzymes but are still easier to digest than more concentrated foods.

The energy that is used to digest food, extract the nutrients, and eliminate the waste products while on a cleansing diet (especially if raw juices and predigested foods are used), is far less than the energy used to digest the everyday, traditional, enzyme-free diet. I have seen tremendous recoveries using predigested food diets that consist of raw sprouts, freshly-squeezed juices, fasting, soaked and sprouted grains and soaked seeds and nuts that are blended before being eaten.

Two major changes take place in predigested foods: (1) the enzyme content sometimes increases 10-fold, and (2) in the predigestion process, the food is broken down into simpler components. Proteins are broken down into amino acids, starches into simpler sugars, and fats into fatty acids. This process relieves the body of breaking down the more concentrated food elements, and it also benefits from the increase in enzymes. The process conserves energy and enzymes for other metabolic functions. In other words, more energy and enzymes can be used in the healing process.

In the detoxification process, what we try to accomplish is to purifiy the blood stream and balance the endocrine glands so that they will work in harmony instead of being overstimulated and exhausted. This purifies the organs and tissues, and takes the stress off the entire body.

The foods that do not get digested properly can cause toxic reactions. For example, Dr. W.W. Oelgoetz had shown that undigested protein, fat, and starch molecules are absorbed into the bloodstream, and when the blood enzyme level is below normal, these obstructions can cause allergies. By giving the enzymes amylase, protease, and lipase orally to his patients, the blood enzyme level was normalized and the allergies were alleviated.[40]

I can personally verify this, having had a history of numerous allergies as a result of living on a traditional Italian diet. I switched to a vegetarian diet and before long, my allergies were gone.

We have discussed that enzymes are a part of our immune system. We have also shown that when there is a lack of lipase in the blood, cholesterol can accumulate. When there is a shortage of amylase, there can be sugar level problems and drastic emotional swings. So there is no need, at this point, to prove that enzymes are used for breakdown of toxins and foreign accumulations in our bodies. It would seem logical then that any diet that would increase the enzyme level would aid the purification process, especially when we now know that enzymes can be reabsorbed and reused in the body.

Many authors have written about the so-called "healing crisis"—myself included. I have discussed this topic thoroughly in my book, *Natural Healing with Herbs,* published by Hohm Press. The "healing crisis" occurs when the system is overloaded with toxins. As a result, the body works to eliminate them through the skin, bowels, sinuses, kidneys, and lungs, producing such symtoms as rashes, lung and sinus congestion, constipation and diarrhea, and urinary tract problems. All of these can be symptoms of body cleansing.

During these crises, enzymes are busy breaking down the accumulations of waste so that the body can rid itself of them. By adding enzymes during the crises, it would seem that it could only result in an improvement and would ultimately aid in the cleansing process. This has proven to be true in nature.

For those who wish to do more research into this topic, you may find great benefits by reading *Food Enzymes for Health and Longevity* and *Enzyme Nutrition* by Dr. Edward Howell.

There are times when detoxification actually weakens the patient, even though a good cleansing diet has been prescribed. The tissues, releasing poisons that have been stored for years, may contain drugs that were ingested years before. If patients have a weak constitution, a chronic disease, or a submissive will, as soon as the first healing crisis is experienced, they tend to go back to their old diet. They think that the cleansing diet isn't worth it because it makes them feel worse, even if only temporarily. These people really are in need of enzyme support.

Taking supplemental enzymes and using predigested foods will take much of the stress from these patients' systems. They also need a good learning process in relation to proper eating habits, fasting, and the *proper* use of colonics and enemas. All of these things are adjuncts to healing.

Bowel cleansing is one of the first things considered by natural healing practitioners. The bowel is the sewage system of the body, and it too must be cleaned out. It has been estimated that 80% of all diseases start in the large bowel. What generally happens is that food that is not digested putrifies in the colon, producing by-products that are reabsorbed through the intestinal tract and deposited in the joints and tissues of the body. It has been shown by Dr. Selle that by adding enzymes to the diet, the fecal bulk is reduced, transit time speeds up, and the nitrogen compounds found in high protein foods that putrify into toxic gases can be reduced from 30 to 60%.[41]

When poisons are eliminated from the tissues, they then enter the blood stream. These poisons cause the endocrine glands to secrete hormones, which in turn affects the eliminative organs, resulting in a feeling of stimulation. When the glands become overworked and exhausted, there will be a feeling of exhaustion and recovery time will be necessary.

The point here is that enzymes can be used not only to maintain health, but also can be used during detoxification programs. Enzymes can be used in both medical and non-medical approaches to healing. They can be the support system to all systems and to health-promotion processes.

There are many enzyme supplements on the market. Some are better than others. Finding the most suitable enzymes, ones that can be substituted for the lack of body enzymes, is of primary importance. Since digestion takes place in the stomach and in the small intestine, enzymes that have activity in both areas will aid the body more than ones that work in just one organ. The stomach and small intestine are where most of the digestion takes place. Both areas have different pHs; consequently, enzymes that can do their digestion in a wider range of pH are much more beneficial.

What is "pH?" "Acid" and "alkaline" are two words used to describe degrees of pH. PH means hydrogen potential. The more hydrogen there is in a solution, the more acidic it is. As the hydrogen concentration decreases, the more alkaline a solution becomes.

So, "acid" and "alkaline," in reality, describe the hydrogen concentration in a solution. Numbers are used to indicate pH ranges. The pH range is from 1 to 14: 1 to 6 is acidic; 1 is very acidic, 6 is only slightly acidic; 7 is neutral; and 8 to 14 indicate that the solution is becoming more and more alkaline.

The body's digestive juices have pH ranges. Protein digestion is partially accomplished in the stomach where hydrochloric acid is secreted along with the digestive juices creating a pH between 1.6 and 4.0. When protein and other food substances are partially digested and become a semi-fluid paste called "chyme," it passes slowly into the small intestine. This chyme has an acid pH. In the first part of the small intestine (duodenum), chyme is neutralized by the pancreatic secretions, which contain bicarbonate ions. This secretion changes the pH from acid to more alkaline (a pH of around 7 to 8). This action is important because pancreatic and intestinal enzymes function best in an alkaline environment.

Pepsin is an enzyme secreted in the stomach that begins digestion of protein foods. It can only function in an acidic digestive juice. When it enters the small intestine, its action is blocked by the neutrality of alkaline pancreatic secretions.

At this point, trypsin, a pancreatic enzyme which also digests protein, is secreted in the small intestine. It more or less takes over where pepsin leaves off. So, the body digests protein in the stomach in an acidic environment, then continues the digestive process in the small intestine in an alkaline environment. The pancreas also secretes amylase and lipase in the small intestine to digest fats and carbohydrates.

For many years, enzymes have been extracted from the intestines and pancreases of animals, put into tablets, and taken orally to aid the digestive process. It has been postulated that enzymes extracted from plants did not have the capabilities of substituting for these tissue enzyme extracts. Dr. Howell and other scientists have proven otherwise.

An interesting research paper was published by the Biochemical Department of E. Merch by Hinnrich, Huffmann, and Lang entitled, "Suitability of the Plant Protease Bromelin for Substitution Therapy in Interestinal Disorders." The paper compared the digestive capabilities of bromelin, an enzyme extracted from pineapples, and the body's own enzymes, pepsin and trypsin.

Dr. W.H. Taylor of Oxford University, investigated the digestive pH of the stomach. He found there to be two pH zones. At the beginning of protein digestion in the enzyme-stomach, the pH ranges from 3.4 to 4.0. As digestion continues, the pH becomes more acidic, ranging from approximately 1.6 to 2.4.

In this research paper, Dr. Taylor stated that the body's enzyme, pepsin, functions best in a pH between 1.5 and 2.5. This means that at the beginning of digestion in the stomach, when the pH is 3 to 4, pepsin is not at its optimum digestive capability. In other words, when the food is in the predigestive stomach, pepsin has little activity. As the stomach becomes more acidic, in approximately $^1/_2$ to 1 hour after food is eaten, pepsin's activity increases.

On the other hand, bromelin was found to be active in a pH range of 3 to 8. It was not only found active in the stomach's higher pH ranges (from 3 to 4), but was also found in the small intestine still actively digesting protein in an alkaline environment (approximately 7 to 8). This again, exemplifies the truth that some enzymes do survive the acid secretions of the stomach.

As previously stated, the pancreas creates an alkaline environment in the duodenum (first part of the small intestine) with its secretions. Also, it secretes its enzyme, trypsin, that continues protein digestion in the small intestine in an alkaline environment.

Bromelin was shown to have the same digestive capabilities as pepsin and trypsin. It has a double effect, both in the stomach and intestines. Thus, it is said that bromelin can be effectively substituted for pepsin and trypsin in enzyme supplementation.

Another misconception is that the only food that gets partially digested in the stomach is protein, leaving fat and carbohydrate digestion up to the pancreatic secretions in the small intestine. It has been established that the enzymes in plants can function in a wide pH range and that they can carry on their activity both in the stomach and small intestine, predigesting starches and fats. The enzymes I speak of are not just proteolytic enzymes, but also fat and carbohydrate digestive enzymes.

Dr. Selle, a physiologist from the University of Texas, fed dogs cereal starch with the addition of the starch digestive enzyme amylase, extracted from barley. The pancreases of the dogs were ligated (tied off) so that no pancreatic digestive juices would affect digestion. The stomachs were then emptied after a certain period of time, and, in some cases, the cereal starch was 65% digested. The barley amylase was proven to digest the starch in the stomach in pHs as acidic as 2.5.

When ptylin (the starch digestive enzyme in human saliva) was compared to this amylase, it was found to be inactivated in the stomach at a 4.5 pH. The ptylin could not function in an acidic pH as well as the plant amylase did. To prove that plant amylase was so effective, one-half hour after it was orally taken, 69-71% of it was found in the small intestine still actively digesting starch.

When the fecal matter was studied for its enzyme content, to see if the plant amylse could be found, there was a larger amount of plant amylase than the body's own pancreatic enzymes.[42]

At this point, I would like to point out that enzymes from fungi (mushrooms and yeast) were used in experiments and they also were found to be very active in both stomach and intestines, and still found to be active in the large intestine.

The starch amylase is not the only enzyme that can function throughout the digestive tract. Plant proteases and lipases can function at a 3 to 8.5 pH, so they can actively continue digestion in both the stomach and small intestine.

Enzymes are needed that can work in the predigestive areas of the stomach, in an acid pH in the lower part of the stomach and in the alkaline environment of the small intestine. Dr. Howell, after years of research, found that the enzymes in certain fungi, if cultured on food materials such as wheat, bran, or soybeans, have the enzymes protease, lipase, amylase, and cellulase. These enzymes can function in tremendously wide pH ranges. They have been found to function throughout the whole intestinal tract, also helping increase deficient blood enzyme patterns. He states that chewing a few of these capsulated enzymes with meals, or by taking the powder in small amounts of water before eating is best. Here is a chart that comapres and demonstrates the ranges of pH in which these enzymes function best:

EFFECTIVE pH RANGES

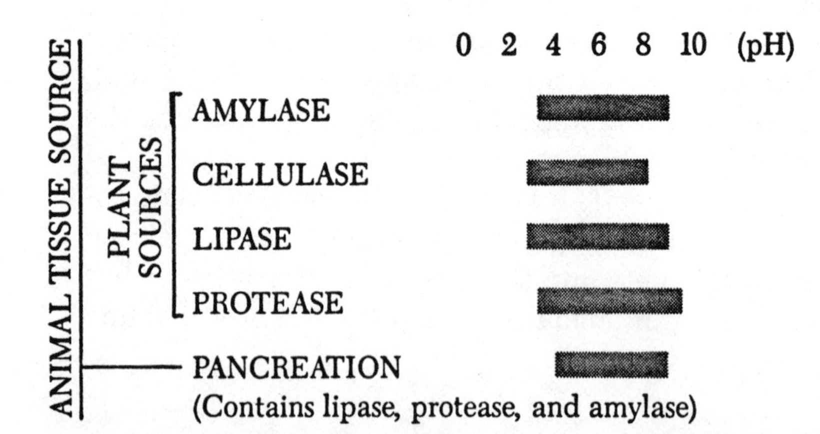

You can see that pancreation cannot function in a wide range of pH. "Pancreation" is a term used to describe the pancreatic digestive secretions containing lipase, protease, and amylase. These enzymes are secreted by human and animal pancreases. Usually they are extracted, purified, and concentrated from slaughterhouse animals.

The fault with them is that they can only function in the small intestine where the pH is slightly alkaline. As supplements, they cannot function in the predigestive stomach where much of the digestion can take place.

Usually, these supplemental pancreatic enzymes (pancreatins) are contained in enteric capsules (capsules treated in such a way as to pass through the stomach unaltered so they can disintegrate in the intestines). These capsules cannot dissolve in the acidity of the stomach. The alkaline juices from the pancreas dissolve the enteric coating and release the enzymes in the small intestine. The enteric-coated capsules do not work at all in the predigestive stomach.

Furthermore, their release is triggered by the pancreatic secretion, which does not take the stress off the pancreas as much as the plant enzymes do. This is because the plant enzymes have predigested a good percentage of the food in the stomach and have begun digesting in the small intestine before the pancreas secretes its enzymes.

Hence, the pancreas secretes fewer of its enzymes because much of the digestion has already taken place. This leaves more enzymes in the body for metabolic functions and body repairs.

The enzymes from certain plants break down and digest food elements through the entire digestive tract. By the time the food gets down into the small intestine, the pancreas has less work to do, and secretes smaller amounts of enzymes. This is a relief to the whole body. These enzymes can be found in most health food stores.

Another point is that when raw foods are eaten, less stomach acid is secreted than when cooked food is eaten. This gives the enzymes contained in the raw food more time for predigestion. When raw food is eaten, the enzymes are attached to their substrate, which protects them from acid digestive juices so that they can function better. If the food is cooked and the enzymes are destroyed, the substrate remains inactive in the predigestive stomach.

If the food substrate is a fat or starch, it must wait until it passes into the small intestine before most of the digestion takes place. Often, when the digestion is sluggish or the enzyme level is low, as found in elderly folks, the food ferments and causes gas, bloating, constipation, colitis, and other problems.

An understanding of these facts permits our logic to tell us that a vegetarian diet with at least 75% raw food is favorable. If this amount of raw food cannot be tolerated, plant enzyme supplements can be used. Even if an all raw food diet is eaten, the proper amount of enzymes may not be guaranteed because citrus fruits and non-starchy vegetables have small amounts of enzymes when compared to starchy, raw foods such as bananas, mangos, and avocados. So, we can see that even the raw foodist can benefit from taking enzymes extracted from plant sources.

CHILDREN AND ENZYMES

As previously pointed out, many diseases in their early stages of development have two major causes. The first is the ingestion of enzyme-deficient foods over a period of time, coupled with eating foods that lack the vitamins and minerals needed daily. Manufacturing procedures, mineral-depleted soils, and cooked, enzyme-free foods create this undernutrition. The second cause is carcinogens, cholesterol, x-rays, drugs, caffeine, and other life-impinging agents used throughout our lifetime. Our children are equally affected by both.

One of the first questions asked is, "How are babies born with health problems?" Complete breast-feeding of infants is very important. Mother's milk has all of the nutrients needed for the growth of the child and a large amount of live enzymes which the baby thrives upon. Milk formulas lack enzymes, and other artificial formulas can be toxic to the child, causing infections, mucus conditions, fevers, diarrhea, colic, and allergies.

Over a period of years at the Infant Welfare Centre of Chicago, the health and development of 20,061 infants was closely monitored for the first nine months of the infants' lives. Of these, 48.5% were wholly breast-fed, 43% partially breast-fed, and 8.5%

wholly artificially fed. The mortality rates of these different groups are shown here:

	No. of Infants	Total Deaths	% of Death of Infants
Wholly Breast-fed	9,749	15	0.15
Partially Breast-fed	8,605	59	.7
Artifically Fed	1,707	144	8.4

The mortality rate among the artificially fed infants was *fifty-six times greater* than among the breast-fed. Four of the 9,749 breast-fed infants died of respiratory infections, compared to eighty-two of the 1,707 artificially-fed infants.

In the United States, one deformed child is born every five minutes, which is the equivalent of one in every ten families. In this country alone, we produce 250,000 deformed infants yearly. Seventy-five percent have mental defects.

There are two major factors involved when considering children's health. The first, of course, is the mother's health. Dr. Bieler, M.D., states that "unless the mother is detoxified before conceiving, the baby comes into the world...full of toxins from the mother's blood and an intestine full of meconium (black bile). The baby is, in fact, so toxic that even with the best care it usually takes three years to eliminate his or her inherited birth poisons."

The second point to be considered here is that once a child has been brought into this world with a weak constitution and a bloodstream laden with poison, he or she is then expected to exist on a diet of concentrated, enzyme-free foods. Heavy starches and mucus-inducing foods (dairy products, grains, sugar, bread) bring on respiratory disorders, asthma, pneumonia, measles, and runny noses. Eating excessively fatty foods brings on high cholesterol problems, acne, and boils.

Just recently (1986), doctors have begun to examine school children for high cholesterol and triglycerides. The foods most school children consume lack the three major enzymes, lipase, protease, and amylase, that help break down foods in the digestive tract. This results in the absorption of large protein and fat molecules which lay the foundation for allergies, obesity, constipation, and fatigue.

There are two major problems that the school systems must cope with: hyperactivity of children and absenteeism. It is very difficult for children to learn valuable information when they cannot concentrate. Read again the section in this book on "Body, Mind, and Enzymes." The relationship between the mind and what a person eats is of paramount importance.

Learning disabilities are often caused by nutrient deficiencies. The majority of deficiencies are caused by foods that lack enzymes—too much cooked food, junk food, processed food, etc. The child's body becomes toxic and the nerves become irritated.

Two children out of every 100 have a neurological disorder. These are disorders that are observed by medical examinations. What about the hyperactive child that doesn't exhibit a true physiological disorder but suffers from an overstimulated nervous system caused by soft drinks, caffeine in chocolate bars, sugar, or other stimulants? Haven't you ever experienced the jitters when you had too much coffee or too much sugar? Consider that a child's body is much more sensitive than our own. In some cases, even just the use of salt can cause hyperactivity, dehydration, and allergies.

To show how important enzymes are, an experiment was performed at Washington University. Surgeons equipped a group of dogs with tubes designed to drain all of the pancreatic juice enzymes out of the body so that the enzymes could not be utilized. Despite the animals' usual access to food and water, profound deterioration set in, and all of them died within a week.

The importance of enzymes cannot be overemphasized. If the lack of enzymes can cause disease, then adding enzymes to the diet, either by supplementation or by eating proper food, will help to prevent disease. Adding enzymes to one's diet is the logical thing to do when we consider that enzymes are used up daily.

Children who maintain high enzyme levels maintain high energy levels. Consider the many ways young adults and growing children lose enzymes. During all fevers and infections, the immune system uses large amounts of enzymes to protect the body and to eliminate toxins and bacteria. In fact, as previously stated, anytime the temperature of the body is raised, enzymes are used up—both during healthy exercise as well as during fever. More calories are burned during exercise, and this natural oxidation process is initiated by enzymes.

Overfeeding a child, especially with cooked, devitalized food, causes the digestive organs to secrete large amounts of enzymes daily, which over a period of time can exhaust the enzyme-producing organs and deplete the immunity of the whole body. This puts a strain on every tissue throughout the body. This child will tend to act exhausted, compared to peers, because his or her energy is being used for digesting foods and coping with the large amounts of waste left over from the unnaturally large food supplies. The body is then required to store large amounts of fat, which puts an additional strain on the heart, kidneys, and lungs.

It should be clear that we use up our given amount of enzymes in many different ways. The proper thing to do is to feed your child and yourself only when you are hungry. This will conserve enzymes. You and your child should eat most of your food raw and add enzyme supplements to enhance digestion. Consider occasional, short fasts. During fasting the enzymes go to work cleaning up undigested materials in the blood and purifying the whole body. Then resume a good healthy diet; cut out the junk food and increase the fresh fruit and vegetables.

THE USE OF ENZYMES IN THE PREVENTION OF ALLERGIES AND CANDIDA

In natural therapeutics, all diseases are considered to be systemic problems. This means that all body processes are involved. To bring about favorable results, the whole body must be influenced in a positive direction.

In considering candida and allergies, there are many similarities between the two, both in cause and prevention. Basically, the symptoms of both maladies are caused by the body's attempt to resist the organisms or toxins that tend to damage tissues and organs. This capacity is called "immunity." In this section, I will discuss basic immune responses of these conditions and how to aid the immune system in preventing such problems.

A major part of our immune system consists of several types of white blood cells, such as lymphocytes, macrophages, T-cells, B-cells, and neutrophils. T-cells get their name because the thymus gland aids in their production. B-cells are named after tiny little sacs in the body called "bursa" where they migrate to be processed until maturity. T- and B-cells are lymphocytes but they differ slightly in their duties.

T-lymphocytes become sensitized to specific toxins or antigens (substances that cause immune responses, such as toxins, drugs, etc.) and attack them whenever they enter the body. B-lymphocytes produce antibodies which are molecules that react with certain antigens in the body aiding in their destruction.

White blood cells help to destroy antigens and other toxins by engulfing them and digesting or partially destroying their substance, making it easier for the body to eliminate them. In most cases, they do this by secreting enzymes that break down the antigens. As previously mentioned, different research experiments that were performed by Dr. Willstatter as far back as 1929 and proved that white blood cells contain eight different types of amylase, protease, and lipase. He states that "white blood cells provide transportation for enzymes throughout the body."

Antigens, bacteria, yeast, and other toxins enter the body through the digestive tract. They leech onto food substances we ingest and then multiply prolifically within the body of the immune system if it is not strong and healthy enough to destroy them. Allergens (substances causing an allergy) can also enter the body simply through the air that we breathe.

Most antigens, bacteria, viruses, and yeasts are proteins. Often the toxins causing allergies and infections are secreted by bacteria that also contain protein substances.

At this point it should be understood that the body needs a tremendous supply of protease (protein digestive enzymes) to counteract the constant bombardment of these proteins to digest and eliminate them. This digestion of protein is done by enzymes, not only in the digestive tract, but in the bloodstream itself. Undigested proteins are often found in the bloodstream. If digestion is not properly accomplished, undigested substances can be absorbed through the digestive tract.

Antigens that cause allergies attach themselves to these proteins in the blood (antigen complex), deposit in the walls of tiny capillaries and secrete substances that cause inflammations which result in swellings, sneezing, hayfever, hives, asthma, etc. In order for the body to rid itself of the allergen, it must be separated from the protein molecule. This is accomplished by enzymes that digest the protein and release the allergen so the body can eliminate it via the lymphatic system. This is why it is so important to keep the lymphatic system clean.

Echinacea is an herb that is a specific lymphatic cleanser. It is an excellent herb to use to combat allergies in combination with enzymes because it stimulates the production of white blood cells, and is used to treat inflammations, lymphatic swellings, and infections.

The importance of understanding why undigested proteins, bacteria, and yeast entering the blood via the intestinal wall have toxic effects on one's system, cannot be stressed strongly enough. Their quick proliferation often leaves the body with numerous symptoms and physical indications.

Candida albicans, which lives almost everywhere in the body, is an example of such a yeast. These yeasts live naturally in the intestinal tract and vaginal areas of animals and humans. They can take over our whole body if the immune system is weakened, such as in advanced cases of AIDS. It is important to realize that yeasts are also protein bodies and can be digested by enzymes if the body has a proper supply.

Candida can change form in the human intestines. It can remain in a yeast-fungal form and enter the circulation or develop a root structure that penetrates the intestines, creating a large enough opening for other bacteria, antigens, and undigested protein to enter. These other substances that enter are a major cause of allergies, anxiety, fatigue, digestive disturbances, vaginitis, cystitis, menstrual problems, and migraine headaches. This is the reason for the similar approach in treating both candida and allergies.

Yeast and most antigens (being protein bodies) can be eliminated by administering supplemental enzymes. These supplements nourish the white blood cells, which causes a direct improvement of the immune system. Some allergies are not protein bodies in themselves. They attach to protein molecules and must be split off from the molecule by proteolytic enzymes and then be eliminated by the lymphatic system.

Undigested proteins, to which yeast and other allergens attach themselves in the circulation, often enter by way of the digestive tract. One way to prevent this is to take plant enzymes with meals to aid in the digestion of these substances.

An effective approach to candida, allergies, or any other systemic problem is to take plant enzymes between meals and thereby help to increase the enzyme activity through the whole body, thus reestablishing these levels both in the digestive tract and the bloodstream.

Again, simply eating larger amounts of raw foods which have the enzymes still present, taking supplemental enzymes, and additionally using echinacea can help alleviate problems of this type. Also, using lactobacillus acidophilus will help check the spread of yeast in the intestines and also nourish the immune system.

SPORTS—ENZYMES—AND NUTRITION

An athlete's main concern should be the type of food he or she eats in order to maintain a healthy body and to replace the nutrients which are lost during exercise or competition.

Enzymes, carbohydrates, proteins, fats, vitamins, and minerals are the fuels that the body needs to function properly. When you exercise, most of these substances are used up rapidly by the body and need replacement.

When I ask individuals if they feel that their diet is correct in relation to the amount of activity they are involved in, their answer is frequently, "Yes, I get plenty of foods high in B vitamins, carbohydrates, proteins, and fats." I have found that these individuals also monitor the nutritional values of foods and calculate their individual needs.

Unfortunately, eating the proper amounts of foods and nutrients is only half the problem. A major concern of an athlete is that his or her system absorbs and properly utilizes the food ingested. "Utilization" is the key word here because the food usually lacks the proper enzymes. Enzymes are essential in the digestion of food and the release of nutrients into the body. Enzymes are present in the blood, muscles, tissues, and organs and are involved in every metabolic function.

Nothing can happen without enzymes. Without enzymes, nutrients cannot be used properly in the body and proteins cannot be digested. This can result in bloating, fatigue, stiffness, and hardening of the arteries. Undigested fats thicken the blood, causing insufficient utilization of oxygen and cholesterol. The consequences of the lack of enzymes is innumerable, but the point to recognize is that enzymes can be a missing link in good nutrition.

A person who exercises regularly is concerned about getting and staying in shape. Strength and endurance are the goals of such individuals. How then can anyone achieve endurance if his blood cells are not getting the proper nutrients?

Nutrients may be present in foods you eat, but the workforce of the body is enzymes. This is why most vitamins are called "coenzymes." This means that they must combine with enzymes before the body can use them.

Over-nutrition and under-absorption result in a low energy system. People often believe that they don't recover from exercise readily enough because they either overdid it or didn't do enough. The problem may actually be that the engine is congested with unusable fuel.

The more enzymes you take in through eating a good quantity of raw foods and supplementing your diet with enzymes, the more energy you will have. It has been said that half the amount of body energy is spent digesting food. If exogenous enzymes (enzymes taken through raw food or through supplementation) are added to the diet daily, more nutrients will be available and less food will be needed, resulting in less digestive stress and waste elimination. This is called "energy conservation." The athlete will be able to work out more often and with greater intensity, and will require less recovery time.

Lou Piccone played ten years of professional football and was a driving force for the Buffalo Bills. He was the only player to master seven different positions on the team. When I met him, he weighed 210 pounds and stood 5' 8" tall. By reducing his meat and dairy consumption and adding plant enzymes to his diet, he lost 30 pounds in two months. We trained together for a three-month period. His diet consisted of some fish and chicken, but primarily fruits and vegetables. My diet consisted of fruits, vegetables, soaked seeds, nuts, and rice. We both supplemented our diets with enzymes at each meal.

Lou then competed in the New York State Masters Track and Field events (also known as the Empire State Games) where he won five gold medals in his athletic events. Usually, when one loses thirty pounds by change of diet while in training, it is reported that one experiences a sense of weakness and a lack of energy. Lou's strength actually remained the same and his recovery time and endurance improved. Both Lou and I were entered in this competition during the summer of 1986—both at age 38. We ran in the 100 and 200-meter races and our finishing times were only a few tenths of a second short of our high school and college track times of almost twenty years ago.

When exercise or sports participation is considered, one must be healthy to carry out the program effectively. We must, out of necessity, add to our enzyme reserve, not deplete it. It is great to enjoy sports and other related activities, but it will be a short-lived experience if our metabolic enzyme reserves are not maintained and we fail to supplement the daily losses to our system.

WHO SHOULD TAKE ENZYMES?

This is a question that only the individual can answer according to his own knowledge and understanding, but certain facts must be considered. Aging correlates perfectly with the enzyme reserve in the body. There is a greater amount of enzymes found in the young person's tissues than in an elderly person's tissues. Taking this into consideration, doing all you can to maintain and increase enzyme levels would be an advantage to tissue and organ longevity.

During all acute and chronic illnesses, enzymes are used up more rapidly than is normal. If one is sick or trying to recover either from an acute or a chronic disease, taking enzyme supplements would be beneficial. People with hypoglycemia, endocrine gland deficiencies, obesity, anorexia nervosa, and stress-related problems, could all benefit from enzyme supplementation.

What about the athlete? He or she may be taking vitamins, minerals, and concentrated foods. What makes all of these elements work? Enzymes. Athletes can benefit by taking enzyme supplements because any time the body temperature is raised, as during exercise, enzymes are used up more rapidly than normal. Carbohydrates are burned more rapidly, and more nutrients are needed for fuel supplies. The athlete usually eats mostly cooked food so it is like burning the candle at both ends. Enzymes are used up rapidly and little is brought in to replenish the supply.

The human machine has an innumerable amount of enzymes. We will be hearing more and more about them in the future. As a person grows more health-conscious, he or she will certainly strive for a greater enjoyment of life. More energy, and a stronger mind and body will be needed to face the stresses that the future will bring. Enzymes can be instrumental in achieving such a goal.

The enzymes that the author is experienced with are manufactured by Foundation Enzyme, Wilmot, Wisconsin and distributed by Rainbow Light, Santa Cruz, California.

JUICING, ENHANCEMENTS, AND ALTERNATIVES

The awareness of drinking freshly-made fruit and vegetable juices daily as a means of supplying much needed vitamins, minerals and enzymes was reemphasized in the 90s by Jay Kordich, better known as the "Juice Man." Jay is a seventy year-old gentleman who saved his own life through dietary changes, particularly in the use of juicing. His example and his writings encouraged Americans to take up the practice of consuming juice, which many did—as much as one to three quarts a day.

As early as the 1940s, and well into the 50s and 60s, Dr. Norman Walker, the guru of juicing, and a naturopathic physician, helped thousands of people regain their health through the use of juicing.

Dr. Max Gerson, M.D., who specialized in therapy for chronic illnesses, also used juicing. He had his patients drink 8 to 10 ounces of freshly-made carrot and apple juice every two hours. He believed that when people were ill, feeding them freshly-made, uncooked food in the form of live juices, was the fastest way to help rejuvenate the body.

Drinking juices puts little stress on the digestive tract since juices are easy to absorb and their nutrients are sent into the bloodstream within minutes. Contrast this to the hours it takes to digest a meal that is eaten, and the small amounts of nutrients that are made available to the body even after several hours of digestion. To make matters worse, most of the foods we eat are already cooked, which destroys 40-60% of the amino acids, a large percentage of the minerals, and all the enzymes if the food is cooked at a temperature over 120°.

Over the years, then, we have been feeding ourselves food that is overcooked and deficient. Besides, the soil the food was grown in was probably already deficient to begin with, and this poor quality of soil reflects itself in the final product. Because juicing is done with fresh, uncooked food, however, many of these deficiencies are avoided, and thus juice can make a remarkable difference for people who are sick. All of a sudden, the body is getting fed with easy-to-digest liquid, food/liquid nutrients, and enzymes. That can make a tremendous difference to our immune systems in particular, as well as our overall health. Using juice is such a safe and effective way to rebalance the body, no matter what its condition. Juices cleanse the system, rebuild it, and balance it.

The use of juices between meals is one of the most effective ways to cut down on cravings for unhealthy or unnecessary foods. Eating several big meals a day makes the our body organisms dull, and diminishes our creativity. Freeing up the energy used for digestion provides us with a greater storehouse of healing potential and intellectual activity. We start to feel nourished by the juice in a way we haven't experienced before. Soon we are choosing to eat in a way that supports this new form of nourishment.

Juice itself is not a therapy, but a foundation for other therapies. Some people can't take heavy supplements or drugs. Others can't radically change their eating habits. But almost everybody can start drinking juice.

In my latest book, *Intuitive Eating* (Prescott, AZ: Hohm Press, 1993), I strongly support the use of raw foods, and emphasize the health benefits of juicing. The intuitive eating system is a gradual program of dietary transition which allows each person to heighten nutritional sensitivity, and to determine which foods are best for them. In three stages, I encourage people to move towards the "ultimate diet," which consists of 70-90% raw foods, including sprouts and juices. Over many years, I have found remarkable benefits, both in weight control and general health, as a result of using this diet myself and recommending it to others.

Fruit juices are the cleansers, and vegetable juices are the essential toners and builders of the body, and in general, I recommend using fruit juice in the morning and vegetable juice in the afternoon and evening.

I further support the use of juices to deal with imbalances in particular systems of the body. By combining the Western nutritional approach (which thinks in terms of eating, digestion, and the distribution of the nutrients through the glands) with the Eastern or Chinese approach (which considers nutrition in terms of energy), I have developed an advanced technology for determining exactly what foods and juices are good for what parts of the body and for what conditions. (See *Intuitive Eating* for a detailed description of my Directional Juicing System.)

The use of raw foods puts people back in touch with the real taste, the real texture, and the whole fiber of foods. Eating raw foods allows us to keep a good balance of enzymes in the system as well. When we combine the use of juicing with the use of raw foods, we get the added advantage of a concentrated dose of vitamins and minerals.

Unfortunately, however, I have found that many people start off their raw food/juicing programs with great enthusiasm but often lose heart, failing to keep up the daily program because they don't want to take the time or energy to prepare their fruits and vegetables for juicing, or because their work schedules make it difficult to have fresh juices every day, or because they travel frequently and thus easily get out of the habit of juicing.

Others fail to engage the juicing aspects of the program because they are initially put off by the cost of a high-quality juice extractor, or the expense of buying fresh produce in large quantities. The commitment to health though food is a commitment that many are unable or unwilling to make.

But I have good news for all health-conscious people—those who are lifelong juicers and want a convenient way to have juice when they work or travel, and those who want the health benefits of juicing without the labor. There are now some excellent powdered juice products out on the market.

For a long time, I have thought that a powdered

juice product with the fiber also added back to it and powdered enzymes added to that would be a superb addition or alternative to juicing. It is now possible, furthermore, to concentrate several different types of juice into one capsule or one teaspoon of powder, thus allowing you to get four to six different vegetables or four to six different types of fruits at one time. This can be particularly advantageous for those who are using a juice combination as a way of dealing with weaknesses in specific body systems.

For enhancing a child's diet, especially for those children who are reluctant to eat fresh or cooked vegetables and fruits, powdered juices can be the answer to a prayer. Imagine giving your children the benefits of a diet of kale, parsley, cabbage, broccoli, apple, barley—all in convenient powdered form.

While I definitely encourage parents to change their own diets so that their children receive the modeling of eating fresh, raw foods, I know this may not be feasible for some, impossible for others. So, the powdered juice alternative is the next best thing.

There are cautions, however. Not all powdered juice products have the same high quality standards. In some situations, companies will use a concentrated juice, which is already cooked with nutrients destroyed, freeze-dry it, and then label it as a powdered juice. This, in my estimation, is a seriously deficient product. So, let the buyer beware!

The products I recommend are ones that are processed in a way that preserves the nutrients and adds fiber and enzymes besides.

ABOUT THE AUTHOR

Mr. Santillo was born and raised in Lockport, New York and received his B.S. degree from Edinboro State Teachers' College in Pennsylvania, attending on football and track scholarships.

Following graduation from Edinboro, Mr. Santillo developed thirty-three different allergies and signs of rheumatoid arthritis. After enduring three years of traditional medical treatment and taking prescription drugs to no avail, Mr. Santillo began his search for a real solution to his problems.

Over many years, he obtained the following degrees: Doctor of Naturopathy (Anglo-American Institute of Drugless Therapy), Health Practitioner (Hippocrates Health Institute of Boston, Massachusetts under Dr. Ann Wigmore and Victor Kulvinskas), Iridology Certificate of Merit (under Dr. Bernard Jensen), and Master Herbalist (from Dr. John R. Christopher at the School of Natural Healing in Utah). Continuing his education, Mr. Santillo has attended many courses and seminars including:

Oriental Herbal Medicine (with Sabhuti Dharmananda, Ph.D., Director of the Institute for Traditional Medicine and Preventative Health Care), Myopractic Therapy (with Dr. Jym Marinakas, N.D.), Medical Botany (from the Platonic Herbal Academy in Santa

Cruz, California, with Ed Smith and Michael Tierra), and eight years of study under Dr. Phil Zimmerman and Dr. Thurman Fleet earning his Doctor's degree from the Concept-Therapy Institute, San Antonio, Texas.

NOTES

1. *Chemical Reviews,* 13 501-12 1933).
2. *Food Enzymes For Health Longevity,* by Dr. Edward Howell—Introduction by Viktoras Kulvinskas, Published by Omangod Press, 1986.
3. *Enzymes & Hormones,* by Elizabeth D. McCarter, B.S., M.Ed., D. Sc., and Robert McCarter, B.S., M. S., Ph.D. Published by Bionomics Health Research Institute of Tucson, 1983.
4. See Note 2 above.
5. *Biochemistry Zeit,* 208: 415-27 (1929).
6. *Journal of Experimental Medicine,* 55: 505-9 (1932).
7. *Enzymologia,* 1: 145-50 (1936).
8. *Compt Rend.,* sociological biology, 112: 549-50 (1933).
9. *Zeit,* physiological chemistry, 221: 13-32 (1933).
10. *Journal of Clinical Investigations,* 13: 517-32 (1934).
11. *American Journal of Digestion, Disease, & Nutrition,* 2: 230-5 (1935).
12. *Proc Soc(iological) (Exp)periments (Bio)logy & (Med)icine,* 28: 948-51 (1930-1).

13. *Enzyme Therapy,* Max Wolf, M.D. and Karl Ransbager, Ph.D. (Regent House, Los Angeles, CA, 1972).
14. *Arch. F. Tierer v. Tierz,* 4: 507-25 (1931).
15. *Klin. Woch,* 9: 2295-6 (1930).
16. *Therap. Arch. U.S.S.R.,* 12: 140-4 (1934).
17. See Note 2 above.
18. *Sei-i-kai Mod. Journal,* 54: 1531-8 (1935).
19. *Arch. Verdanugskrankh,* 49: 168-200 (1931).
20. *Biochem. Zeit.,* 240: 328-56 (1931).
21. *American Journal of Digestive Diseases,* 5: 184-9 (1938).
22. *Journal of Experimental Zoology,* 76: 325-52 (1937).
23. *The McDougall Plan,* by John A. McDougall, M.D. & Mary A. McDougall (New Century Publishers, Inc., N.J. (1983).
24. *Ziet, Physiol. Chem.,* 221: 13-32 (1933).
25. *Bibliotek for Larger,* 123: 437-70 (1931).
26. *Nagoya Journal of Medical Science,* 3: 51-73 (1928).
27. *Biochem Zeit,* 275: 216-33 (1935).
28. "The Influence of Food Cooking on the Blood Formula of Man" by Paul Kouchakoff, M.D. (Institute of Clinical Chemistry, Lausanne, Switzerland, 1930).
29. *Enzyme Nutrition* by Dr. Edward Howell.
30. *Ibid.*
31. *Ibid.*
32. *Ibid.*
33. Same as Note 22 above.
34. Same as Note 28 above

35. Same as Note 28 above.
36. *Survival into the 21st Century* by Viktoras Kulvinskas, (Omangod Press, Fairfield, Iowa, 1975).
37. *Ibid.*
38. *Biochem. Zeit,* 288:149-54 (1936).
39. *American Journal of Digestive Diseases & Nutrition,* (Jan. 1934-5).
40. Same as Note 2 above.
41. *Journal of Nutrition,* 13: 15-28 (1937).
42. *Ibid.,* 12: 59-83 (1936).

RETAIL ORDER FORM

NAME: ______________________________

SHIPPING ADDRESS: ______________________________

CITY: ____________________ STATE: __________ ZIP CODE: __________

QTY.	DESCRIPTION	COST W/SHIPPING	TOTALS
	NATURAL HEALING WITH HERBS, 408 pages by Humbart "Smokey" Santillo, N.D.	$18.00	
	HERBS, NUTRITION AND HEALING, 330 minutes 4-Cassette Herbal Seminar Series by Humbart "Smokey" Santillo, N.D.	43.00	
	FOOD ENZYMES: The Missing Link to Radiant Health, 108 pages by Humbart "Smokey" Santillo, N.D.	9.25	
	FOOD ENZYMES: The Missing Link to Radiant Health, 210 minutes 2-Cassette Recording, Plus a Bonus 1-Hour Seminar and a Copy of the Book	43.00	
	ENERGETICS OF JUICING: The Key to Longevity, Health & Energy, 120 minutes 2-Cassette Seminar and a 48-Page Book by Humbart "Smokey" Santillo, N.D.	31.75	
	INTUITIVE EATING: EveryBody's Natural Guide to Total Health and Lifegiving Vitality through Food, 450 pages by Humbart "Smokey" Santillo, N.D.	20.00	
		TOTAL ENCLOSED U.S. FUNDS ONLY	

METHOD OF PAYMENT:

☐ Check or M.O. payable to Hohm Press

☐ Visa ☐ MasterCard

Card #______ – ______ – ______ – ______ Exp.______

Signature______________________________

For shipping outside the U.S., please add 25% for U.S. Mail delivery.

Prices subject to change over time.

Mail to: HOHM PRESS • P.O. Box 2501 • Prescott, AZ U.S.A. 86302

INTUITIVE EATING:

Reclaiming What Your Body Knows and Needs

by Humbart "Smokey" Santillo, N.D.

"Imagine being so attuned to your body that you knew exactly what it needed—when to add protein to your diet; what vegetables and fruits would immediately remedy any imbalance; what foods would help you through a period of emotional turmoil. This is no fantasy, however. It is a very real possibility. One that I have lived, and shared with thousands of clients. This is INTUITIVE EATING."

—*"Smokey" Santillo, from Chapter 1*

The natural voice of the body has been drowned out by the shouts of addictions, overconsumption, and devitalized and preserved foods. Millions battle the scale daily, experimenting with diets and nutritional programs, only to find their victories short-lived at best, confusing and demoralizing at worst. Intuitive Eating is an alternative.

Intuitive Eating is a handbook for developing a personalized relationship to food and nutrition characterized by self-awareness and environmental sensitivity. The author lays a firm foundation in the basics of nutrition and then guides the reader through a gentle, three-stage transitional diet with practical directions in the use of fresh juices and enzymes.

Meal plans and sample recipes, including healthy snacks and desserts, make the transition easy and enjoyable. Extensive charts provide an invaluable reference on the nutritional contents (protein values, vitamins, minerals, fat content, etc.) of most common foods.

Preface by Victor Kulvinskas, author of *Survival Into The 21st Century*. Introduction by Jay Kordich, author of *The Juiceman's Power of Juicing*.

INTUITIVE EATING: *Reclaiming What Your Body Knows and Needs*
by Humbart "Smokey" Santillo, N.D., $16.95 US Funds
• 0-934252-27-0 Health/Nutrition, 420 pages/Trade Paper

HERBAL SEMINAR TAPES

These tapes are a live recording of an Herbal Seminar presented to laypersons, doctors, and other professionals by Humbart "Smokey" Santillo, N.D. The presentation begins with the basics of herbology and extends into a comprehensive coverage of therapies used in clinics throughout the world. It correlates historical herbology with the most recent in American and Chinese herbal philosophies. Following is a description of the content of each tape:

TAPE #1

Side A—This side explains the role that the mind plays in healing—how ideas and thoughts affect the body and how, when negative, they result in dis-ease (disease). An interesting discussion explains how parts of the brain relay thought energy throughout the body, causing tension, stress, and other psychosomatic disorders. Once you understand this material, you are in a position to improve your mental attitude and avoid the negative effects of stress. The physical cause of disease is explained on this side along with other related subjects such as acid-alkaline balance, cleansing diets, detoxification, enervation (lowered nerve energy), and toxemia. These subjects lead into a complete discussion of acute and chronic diseases.

Side B—The three functions of herbs: eliminating, maintaining (balance), and building, are presented. These functions are coordinated with the healing purposes of herbs, such as demulcents, laxatives, expectorants, etc. Also discussed here, and continuing to Tape #2, are eight major therapies used to treat disease.

TAPE #2

Side A—This tape explains treatments such as stimulation therapy, tranquilization therapy, blood purification and tonification therapies.

Side B—This side continues with more therapies such as sweating, emesis, diuresis, and purging. Herbs are listed under each therapy with a full discussion of herbal properties and their definitions.

TAPE #3

Side A—This tape gives a complete explanation of how to prepare and use herbal infusions, decoctions, boluses, douches, electuaries, pills, enemas, tinctures, and oils.

Side B—A continuation of the previous side with discussions of how to prepare and use poultices, plasters, castor oil packs (for treatment of tumors), capsules, salves, and concentrates.

TAPE #4

Side A—In Chinese medicine, diseases are categorized as being either hot or cold. This concept is explored along with suggested ways for determining which illnesses fall into these categories. Dietary suggestions are presented, including cleansing and transition diets. Primary and secondary symptoms are discussed, along with how to determine one's major weaknesses and strengths. This gives the herbalist a basis to determine which part of the body to treat.

Side B—This side includes a thorough explanation of how to treat colds, fevers, and flus using cold sheet treatments, hydrotherapy, and the herbs fenugreek, catnip, fennel, comfrey, and thyme. Most important on this tape is an explanation of how to build proper herbal formulas for both acute and chronic illnesses. The final part of this presentation deals with how each disease goes through five stages of development from the onset to the cure. Symptoms are discussed to help a therapist recognize the stages of the disease and choose the proper herbs for treatment.

* * *

These tapes include concentrated, easy-to-understand information which can be of assistance to anyone who has the interest. This information is invaluable for laypersons, doctors, or students who want to enhance their own health or wish to help others.